ELECTRICITY AND MAGNETISM

ELECTRICITY AND MAGNETISM

ERIC GERARD

Translated from the fourth French edition under the
supervision of Dr. Louis Duncan,

by

R.C. DUNCAN

MAVEN BOOKS

Chennai   Trichy   Tirunelveli   New Delhi

MAVEN BOOKS

An Imprint of **MJP Publishers**

ISBN 978-93-88191-53-1   **MAVEN Books**

All rights reserved   No. 44, Nallathambi Street,
Printed and bound in India   Triplicane, Chennai 600 005

MJP   705   © Publishers, 2018

Publisher :   **C. Janarthanan**

PUBLISHER'S NOTE

The legacy of a country is in its varied cultural heritage, historical literature, developments in the field of economy and science. The top nations in the world are competing in the field of science, economy and literature. This vast legacy has to be conserved and documented so that it can be bestowed to the future generation. The knowledge of this legacy is slowly getting perished in the present generation due to lack of documentation.

Keeping this in mind, the concern with retrospective acquiring of rare books has been accented recently by the burgeoning reprint industry. MAVEN Books is gratified to retrieve the rare collections with a view to bring back those books that were landmarks in their time.

In this effort, a series of rare books would be republished under the banner, "MAVEN Books". The books in the reprint series have been carefully selected for their contemporary usefulness as well as their historical importance within the intellectual. We reconstruct the book with slight enhancements made for better presentation, without affecting the contents of the original edition.

Most of the works selected for republishing covers a huge range of subjects, from history to anthropology. We believe this reprint edition will be a service to the numerous researchers and practitioners active in this fascinating field. We allow readers to experience the wonder of peering into a scholarly work of the highest order and seminal significance.

MAVEN Books

CONTENTS.

Chapter I. Introduction.

Chapter II. Magnetism.

INDUCED MAGNETIZATION.

CHAPTER V. Electromagnetism.

MAGNETIC PHENOMENA DUE TO CURRENTS.

APPLICATIONS RELATING TO THE MAGNETIC POTENTIAL OF THE CURRENT.

ELECTROMAGNETIC ROTATIONS AND DISPLACEMENTS.

ELECTROMAGNETS.

CHAPTER VI. Units and Dimensions.

ELECTRICAL AND MAGNETIC UNITS.

CHAPTER VII. Electromagnetic Induction.

CHAPTER IX. The Propagation of Currents.

CHAPTER X. Electrical Measurements.

MEASUREMENT OF THE INTENSITY OF A MAGNETIC FIELD.

MEASUREMENT OF MAGNETIC PERMEABILITY.

MEASUREMENT OF COEFFICIENTS OF INDUCTION.

THERMO-ELECTRIC COUPLES.

INTRODUCTION.

UNITS OF MEASUREMENT.

A phenomenon is well known only when it is possible to express it in numbers. (KELVIN.)

1. Fundamental Units.—All electrical actions are referred to forces, and are consequently expressed by the aid of the three fundamental quantities, length, mass, and time.

To measure these quantities electricians have chosen the *centimetre*, the *gram*, and the *second* as units.

The centimetre is approximately the billionth part of the terrestrial quadrant; rigorously speaking, it is the hundredth part of the standard metre measured by Delambre and Borda, and kept at the international conservatory at Sèvres.

The gram represents about the mass of a cubic centimetre of distilled water at its maximum density. There is also a standard kilogram at Sèvres.

The second is the 86,400th part of the mean solar day.

These units, called *fundamental*, are represented by the symbols $[L]$, $[M]$, $[T]$.

The *numerical value* of a quantity is expressed by its ratio to the unit chosen. A length measured by a number l will have a concrete value equal to $l[L]$. If we adopt another unit $[L']$, there will be a numerical value l', such that

$$l'[L'] = l[L], \quad \text{whence} \quad \frac{l}{l'} = \frac{[L']}{[L]}.$$

We see, therefore, that the numerical value of a quantity is in inverse ratio to the magnitude of the unit chosen.

2. Derived Units.—In order to express the various physical quantities, arbitrary units might be chosen quite independent of each other. This method, which was long followed, presents no inconvenience when the measures are *relative*, that is, when they are directly compared with their units. But more often quantities are measured by units of other kinds, making use of the inter-relations between the different quantities. Such a system of measurement is called *absolute*. For example, to measure a surface we do not compare it directly with a standard area, but determine its linear elements, by aid of the unit of length, and then apply the relation existing between an area and its linear dimensions.

For a square s, with a side l, the relation is $s = kl^2$. If $l = 1$, $s = k$.

The arbitrary factor k, which represents the area of a square having unit side, may be put equal to 1. The unit of area thus determined is the square whose sides are equal to one centimetre; it is connected with one of the fundamental units, and for this reason is called the *derived unit* of area.

In the same way the derived unit of volume is the cube having sides of one centimetre.

We can thus define derived units for all the physical magnitudes, getting rid of the arbitrary coefficients in the relations which unite these magnitudes together.

The *system of units* determined in this way is called by the initials of the fundamental units chosen.

The C. G. S. system of units, adopted by electricians, has as its basis the centimetre, the gram, and the second.

3. Example of a Derived Unit.—The velocity of a moving body, traversing a path l in a time t, is given by the equation

$$v = k\frac{l}{t}.$$

k expresses the velocity of a moving body traversing unit length in unit time. This velocity is chosen as unity, thus eliminating a factor inconvenient in calculation. This unit is the velocity of one *centimetre per second*; it may be expressed in symbolic form

$$[LT^{-1}].$$

4· Dimensions of a Derived Unit.—Such an expression, which shows the dependence of the derived unit on the fundamental units, exhibits the *dimensions of the derived unit*. It enables us to follow the variation of the derived unit when the fundamental units are changed. If, for example, we measure the time in hours and the length in metres, the derived unit of velocity will be $[L'T'^{-1}] = 100 \times 3600^{-1}[LT^{-1}]$, or the thirty-sixth part of the unit defined above.

Every relation between physical quantities is independent of the units chosen to measure it, therefore this relation must be homogeneous with regard to the fundamental units.

Thus the equation $v = al$, in which v represents a velocity, l a length, and a an abstract number, is inconsistent, for only the first member would vary with the unit of time.

5. Mechanical Derived Units.—Following the line of reasoning used above, it is readily seen that the *unit of angular velocity* is the velocity of a moving body which passes over unit angle in unit time. As the unit angle or *radian* (arc equal to the radius) is defined by a simple numerical ratio, the dimensions of angular velocity reduce to $[T^{-1}]$.

The *unit of acceleration* is the acceleration by which the velocity is increased by one unit per second. Dimensions $[LT^{-2}]$.

The unit of *quantity of movement* (*momentum*) is the momentum of unit mass moving with unit velocity. Dimensions $[LMT^{-1}]$.

The *unit of force*, which has received the name *dyne*, is the force which, applied to unit mass, impresses upon it unit acceleration. Dimensions $[LMT^{-2}]$.

The ordinary unit of force is the *weight of the gram*, that is, the force capable, in our latitude,* of impressing on unit mass an acceleration approximately equal to 981 cm. per second. The gram is consequently equal to 981 dynes.

The *unit of work*, called *erg*, is the work done by unit force on a body moving in the direction of its action over unit length. Dimensions $[L^2MT^{-2}]$.

The ordinary unit of work is the kilogrammetre, which is equal to 981×10^5 ergs, in our latitude.*

The *unit of power*, or *erg per second*, is the power developed when unit work is done in unit time. Dimensions $[L^2MT^{-3}]$.

The *ordinary units of power* are the *cheval-vapeur* (French horse-power), which is equal to $75 \times 981 \times 10^5 = 736 \times 10^7$ ergs per second, and the *poncelet*, defined by the Congress in Mechanics, 1889, as equivalent to 100 kilogrammetres per second, or 981×10^7 ergs per second.

The *unit of density* is the density of a body which contains unit mass in unit volume. Dimensions $[L^{-3}M]$.

The *unit of modulus of elasticity* is the modulus of a body which, supporting unit force per unit of section, receives

---

* Liège, lat. 50° 45' approx.

an elongation equal to its original length.   Dimensions $[L^{-1}MT^{-2}]$.

6. Principle of the Conservation of Energy.—Work applied to a system is capable of various effects.   It may be used : (1) To increase the active energy of the masses, or, to use Rankine's expression, to develop *kinetic energy*, represented by the product of half the sum of the masses into the square of their velocity.   (2) To overcome the friction of the system ; it was long believed that this effect represented a loss of energy, but thermodynamics has shown that in such a case there is generated an amount of heat equivalent to the work expended.   (3) To overcome molecular forces, such as elasticity, chemical affinity ; or to overcome natural forces, such as gravitation, magnetic attraction, etc.   In this case the work is stored up in the system in the form of *potential energy*, which is again transformed into kinetic energy or heat, when the system is abandoned to the reaction of the forces concerned.

Let us suppose, for example, that we raise a weight or stretch a spring which sets a clockwork in motion.   The potential energy given to the weight or spring is transformed into kinetic energy when the mechanism is allowed to operate, and this kinetic energy is itself reduced to heat by the friction of the wheelwork.

The tendency of modern science is to refer these diverse varieties of energy to a single one, kinetic energy ; calorific, luminous, or electrical radiations, for example, which to us seem potential forms of energy, might be reduced to special modes of motion of the *ether*.

The study of physical phenomena has given us a natural law of the highest importance.   *The energy of a system is a quantity which cannot be either increased or diminished by any mutual action between the bodies which compose the system.*

This law of the *conservation of energy*, together with that of the *conservation of matter*, rules supreme in physical science.

From this principle it results that a system cannot of itself produce more than a limited quantity of external work, whence the impossibility of perpetual motion.

The persistence and the indestructibility of *energy* make it as much a physical entity as *matter* is, and give it a leading place among the magnitudes considered in mechanics. Energy assumes indifferently a mechanical, electrical, thermic, or chemical form. Experiment shows that the two former are capable of being entirely transformed into one of the two latter, but that only a portion of thermic or chemical energy can be made to assume a mechanical or electrical form.

In whatever form energy may be, it possesses a mechanical equivalent; it is therefore homogeneous with work $[L^2MT^{-2}]$, and may be measured in mechanical units. It follows that the C. G. S. unit of heat equals the erg.

One *gram-degree*, or *lesser calorie (caloriegram)*, represents 4.2×10^7 C. G. S. units of heat.

The electrician has constantly occasion to apply the principle of the conservation of energy, which we have just defined, for the essential rôle of electricity is to serve as agent for the transformation of energy. The energy of the electric current is produced from the work done by chemical affinity in batteries, by using up heat in thermo-electric couples, or by an absorption of mechanical power in dynamos.

The energy of the current is, in its turn, transformed into heat and light in the conductors and electric lamps; it is capable of decomposing an electrolyte or of overcoming the resistance offered to the motion of an electromotor.

The marvellous facility with which electricity lends itself to the transmission and transformation of energy, and which justifies the increasing number of applications of this agent, leads the electrician to compare phenomena of very diverse kinds, the measurement of which demands such a system as the C. G. S., embracing all physical magnitudes.

7. Multiples and Submultiples of the Units.—The use of the units just described leads sometimes to very large or very small numerical values. By way of abbreviation we make use of multiples or submultiples designated by such prefixes as kilo-, mega- (one million), milli-, micro- (one millionth).

$$\text{Thus, one megadyne} = 10^6 \text{ dynes;}$$
$$\text{one microdyne} = 10^{-6} \text{ dynes.}$$

8. Application of the Dimensions of Units.—The dimensions of the units are of use not only in verifying the homogeneity of formulæ, but they allow us, as Bertrand has shown, to predict the form of a function when the physical quantities which enter into it are known. Suppose, for example, that experiment has shown that the velocity of propagation of an undulatory movement in a medium depends on the modulus of elasticity and the density of the medium.

Then the velocity v is a function of the elasticity e, and the density d;

$$v = \phi(e, d).$$

If we consider the dimensions of the quantities which enter into this equation, we have

$$vLT^{-1} = \phi(e\, L^{-1}MT^{-2},\ dL^{-3}M).$$

As M is wanting in the first member, the homogeneity of the function requires it to be eliminated from the second, which is obtained by adopting the form

$$vLT^{-1} = \phi\left(\frac{e\,L^{-1}MT^{-2}}{d\,L^{-3}M}\right) = \phi\left(\frac{e}{d}L^2T^{-2}\right).$$

To bring L and T to the same degree in both members, it is clear that the function must be a radical of the second degree. From what precedes we conclude that v is a linear function of

$$\sqrt{\frac{e}{d}}.$$

And in fact experiment shows that the relation sought is

$$v = \sqrt{\frac{e}{d}}.$$

GENERAL THEOREMS RELATIVE TO CENTRAL FORCES.

9. Definitions.—Forces are called *central* whose direction passes through definite points, called centres of force, and whose intensity is a function of the distance between those points.

The *Newtonian central forces*, such as gravitation, electric and magnetic attraction, are inversely proportional to the square of the distance between the acting centres.

In studying the effects of these forces, it is a matter of indifference whether they emanate from the centres themselves or have their seat in the medium which separates these centres. Thus, to account for the universal attraction of matter, the simplest way is to assume that the attractive force is a property of all ponderable bodies, which act upon each other at a distance. This hypothesis has the advantage of lending itself readily to calculation. It has sufficed as a basis for celestial mechanics.

Nevertheless it does not satisfy the intellect. The ordinary methods used in the transmission of forces show us the necessity of an intermediary, such as a tense cord, air or water under pressure, and this permits us at least to limit to intermolecular space the idea of action at a distance. Again, the direct action of one body on another takes for granted an instantaneous effect. Now physical phenomena, even the most rapid, have a finite time of propagation.

To account for observed phenomena, physicists have been led to suppose the universe filled with an ocean of *ether*, whose waves, representing heat, light, and electrical energy, are propagated with a velocity of 3×10^{10} centimetres per second, so that they take about eight minutes to reach us from the sun.

However, for simplicity of treatment, we shall admit *provisionally* that central forces are due to the bodies from which they seem to emanate, or to an *agent* diffused through these bodies.

In the case of *gravity* the observed actions are attributed to the *mass* of the body.

In the case of the electrical phenomena that are manifested between bodies that have been rubbed, we shall say that an *agent*, called *electricity*, has been developed on these bodies, and, without making any supposition as to its nature, we shall speak of *quantity*, *mass*, or *charge* of the agent, these terms expressing merely a factor proportional to the effects produced.

Thus we shall say that two bodies possess equal *quantities* of the agent when they produce equal effects on a third body. The quantities of the agent will be doubled, or tripled, when the forces developed are double or triple.

The quantity of agent per unit of area or per unit of volume is called, the *surface density* or *volume density*.

10. Elementary Law Governing the Newtonian Forces. —The preceding definitions amount to saying that the force exerted between two quantities of the agent is proportional to the product of these quantities, since it is proportional to each one of them. It is also a function of the distance between the masses concerned. In the case of Newtonian forces it is inversely proportional to the square of the distance.

If, then, we express by m, m' two quantities of the agent, and by l their distance, the force

$$f = k\,\frac{mm'}{l^2}.$$

The action exerted on one of the masses considered as unity would be expressed by

$$H = k\,\frac{m}{l^2}.$$

In the case of electric and magnetic actions, masses of the same nature repel each other, contrary to what holds in case of gravitation.

In a logical system of units, the constant k is not a simple numerical factor. Consider the attraction of heavy bodies and replace force, mass, and distance by their dimensions: then the condition of homogeneity demands that k have dimensions $[L^3 M^{-1} T^{-2}]$.

11. Field of Force.—Let us suppose that the quantities m, m', m'' of the agent are concentrated in physical points, occupying given positions in space. If we bring into their vicinity a mass of the agent equal to unity, it is acted upon by the forces emanating from m, m', m'', which form a resultant having a definite direction and intensity. By changing the position of the point charged with unit mass of the agent we can obtain the intensity and sign of the resultant force for every point in space.

The space in which such forces are manifested is called a *field of force*, and the resultant force, just defined, is the *intensity of the field* at the point where the unit mass is placed. The direction of the resultant is called the *direction of the field*.

The value of the intensity of a field is directly deduced by the application of the elementary law; but the necessary calculations by this method become extremely complicated in the case of a number of acting masses, since the elements to be combined are *vectors*, that is, quantities having given magnitudes and directions, and combining according to the parallelogram of forces. The procedure is especially involved when the analytical method is employed.

The solution of the problem is reduced to a simple algebraic addition followed by a differentiation, by taking into consideration a new function defined by Laplace and investigated by Gauss and Green under the name *potential*.

12. Potential.—Let us suppose that unit mass is dis-

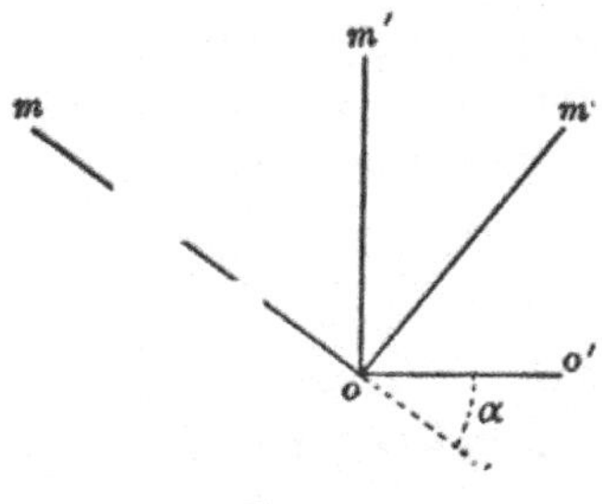

Fig. 1.

placed in the field by an infinitesimal distance, under the action of the forces acting upon it.

Let $oo' = dr$ be the displacement. The force due to the mass m is $\dfrac{km}{l^2}$, and the work done under the action of this force is

$$\frac{km}{l^2}\, dr \cos \alpha = \frac{km}{l^2} dl.$$

Likewise the work done by m' is

$$\frac{km'}{l'^2}\mathrm{d}l';$$

by m'',

$$\frac{km''}{l''^2}\mathrm{d}l''.$$

These single expressions for the work done are to be added, since they are taken in the same direction; the total work is therefore expressed by

$$k\Sigma\frac{m\mathrm{d}l}{l^2}.$$

This sum is the differential of the function

$$-k\Sigma\frac{m}{l}+\text{const.}$$

The expression $+k\Sigma\frac{m}{l}$, whose differential, taken with the contrary sign, represents the elementary work of the forces of the field, has been given the name of *potential* by Gauss. We shall designate it by the letter U:

$$U=+k\Sigma\frac{m}{l}.$$

For a point in space, therefore, the potential is proportional to the sum of the ratios of the acting masses to their distances from the point.

The potential permits us readily to define the work accomplished by the forces of the field.

Thus, if we integrate the expression

$$k\Sigma\frac{m\mathrm{d}l}{l^2}=-\mathrm{d}U$$

between two positions O_1, O_2, occupied by the unit mass, we get

$$k\Sigma m\int_{O_1}^{O_2}\frac{\mathrm{d}l}{l^2}=\int_{O_1}^{O_2}-\mathrm{d}U=U_1-U_2.$$

The work done by the field on unit mass displaced from the point O_1 to O_2 is equal to the difference of the values of the function U at the two points.

The *work* depends solely upon the position of the initial and final points, and not on the path followed by the unit mass between these points.

If the unit mass should pass from the point O_1 to an infinite distance from the acting masses, we would have   .

$$k \Sigma m \int_{O_1}^{\infty} \frac{\mathrm{d}l}{l^2} = \int_{O_1}^{\infty} - \mathrm{d}U = U_1.$$

Hence we see that *the potential at any point is measured by the work done by the field in displacing unit mass from the given point to an infinite distance from the acting masses, that is, to the limit of the field.*

The potential function furnishes a simple expression for the intensity of the field.

Let H be the component of intensity in a direction l. The elementary work Hdl is likewise expressed by the differential, taken with contrary sign, of the potential in this direction:

$$H\mathrm{d}l = - \frac{\mathrm{d}U}{\mathrm{d}l}\mathrm{d}l;$$

whence

$$H = - \frac{\mathrm{d}U}{\mathrm{d}l}.$$

The component of field intensity in a given direction is expressed by the derivative, taken with contrary sign, of the potential in that direction.

The force is directed towards the points where the potential diminishes.

13. Equipotential Surfaces.—Put

$$U = \phi(x, y, z) = \text{constant},$$

$x, y,$ and z representing the points of the field by rectangular co-ordinates.

This equation represents a surface at every point of which the potential has the same value. Consequently the forces of the field have a zero resultant along this surface, the normal to which represents the direction of the field at each point.

The surfaces thus defined are called *equipotential surfaces* or *level surfaces*, by analogy with the free surface of a liquid, everywhere normal to the force of gravity.

Designating by n a direction normal to the equipotential surface, the field intensity in a point of the surface is expressed by

$$H = -\frac{dU}{dn}.$$

We can get a representation of the distribution of the forces of the field by imagining in the field a series of similar surfaces sufficiently near to each other and corresponding to potentials which increase in arithmetical progression.

A mass free to move in the field will follow a path cutting the equipotential surfaces perpendicularly. This curve, whose tangent represents in each point the direction of the field, has been named by Faraday a *line of force.*

The field intensity is obviously in inverse ratio to the segment of line of force comprised between two consecutive equipotential surfaces.

14. Case of a Single Mass.—The case of a single acting

mass gives an example of a field easily defined. The equipotential surfaces are concentric spheres, whose radii represent the lines of force.

Let us suppose a mass m, such that $km = 6$, concentrated in a point A.

The concentric circles represent the intersection by a plane passing through the mass m of the equipotential surfaces 1, 2, 3, 4, 5, 6. The radii of these circumferences are respectively 6/1, 6/2, 6/3, 6/4, 6/5, 6/6.

15. Uniform Field.—We see that as the potential decreases the equipotential surfaces are successively further and further apart. At a sufficiently great distance from the

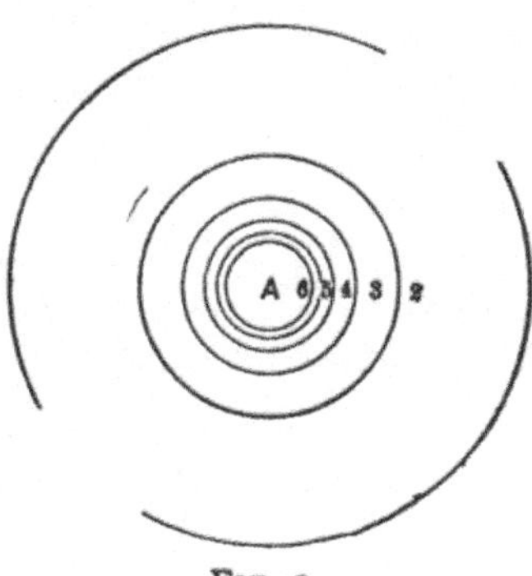

Fig. 2.

centre the lines of force drawn through a region of small extent are practically parallel, and the equipotential surfaces are comparable to planes in this region. In the case of gravitation, for example, no appreciable error is caused by taking, in the space occupied by a laboratory, the verticals as parallel.

A field represented in this manner by equipotential planes and lines of force perpendicular to them, whose intensity is constant in magnitude and direction, is called a *uniform field*.

16. Case of Two Acting Masses.—Let us consider the

case of two acting masses, such that for one of them $km = 20$, and for the other $km' = 5$.

To determine the intersection of the equipotential surfaces due to the two centres by a plane passing through these centres, commence by tracing the circular equipotential lines due to each centre considered separately.

Let n_1, n_2, n_3, n_4 . . . . be the circles drawn around the first, and n_1', n_2', n_3' . . . those enveloping the second.

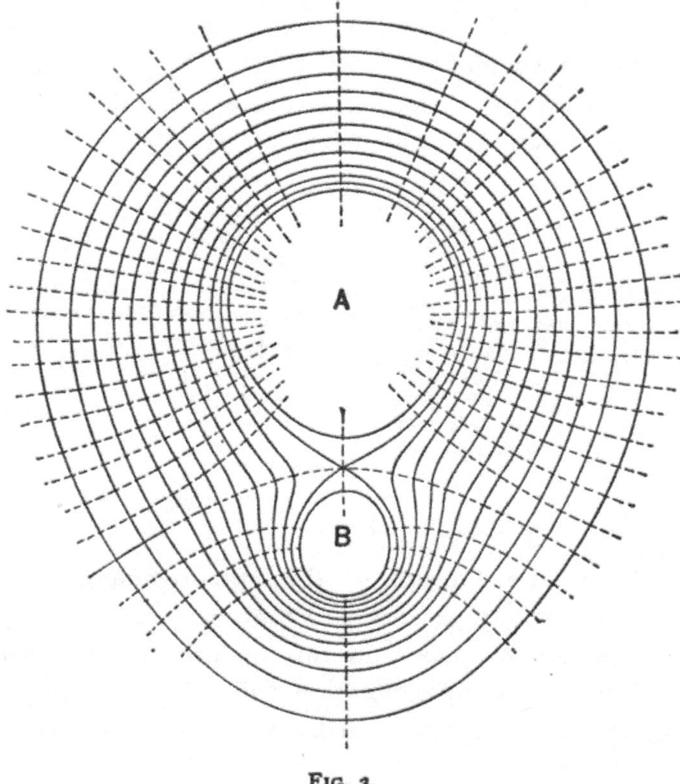

FIG. 3.

The equipotential line of the order 5 will evidently pass through the intersections of the circumferences n_4, n_1' ; n_3, n_2' ; n_2, n_3' ; n_1, n_4'.

The equipotential line of the order 4 will pass through

the intersections of the circumferences n_3, n_1'; n_2, n_2'; n_1, n_3', and so on for the other orders.

The lines of force will be curves normal to the equipotential lines obtained.

In Fig. 3, taken from Maxwell's *Electricity and Magnetism*, the two masses above mentioned are concentrated in the points A and B. The full lines are equipotential; the lines of force are shown by dotted lines.

17. Tubes of Force.—Trace any closed curve in a field, and imagine that a line of force passes through each point of this curve. All these lines taken together form a tubular surface, called a tube of force. In the case of a single centre of force the tubes of force are conical. In a uniform field they are cylindrical.

18. Flux of Force.—The intensity of any field is constant over an infinitely small surface ds. The product of this surface into the component of the intensity normal to the surface is called the *flux of force across the surface*.

Let α be the angle of the direction of the field with the normal, the flux of force will be represented by

$$\mathrm{d}N = H \cos \alpha \, \mathrm{d}s.$$

The flux of force across a finite surface is given by

$$N = \int H \cos \alpha \, \mathrm{d}s,$$

the integration being extended to every element of the surface under consideration.

In the case of a closed surface the flux is said to be *issuing* when the lines of force are directed towards the exterior of the surface, and *entering* in the opposite case.

By considering the angle α to be made by the direction of the field with the normal *exterior* to the surface, the change of sign of $\cos \alpha$ allows us to distinguish the issuing from the entering flux.

19. Theorem.—*The flux of force which traverses a tube of infinitely small section is independent of the inclination of the section to the axis of the tube.*

Thus, $$dH = H \cos \alpha \, ds = H d\sigma,$$

$d\sigma$ representing the section of the tube normal to the axis, and H the intensity of the field at this point.

20. Gauss' Theorem.—Between the masses of a field and the flux traversing a surface which envelops these masses there exists a simple relation, very frequently used, as follows:

The flux of force traversing a closed surface in a field is equal to $4\pi k$ times the sum of the masses enveloped by this surface.

I. Let us first consider a single mass m, concentrated in

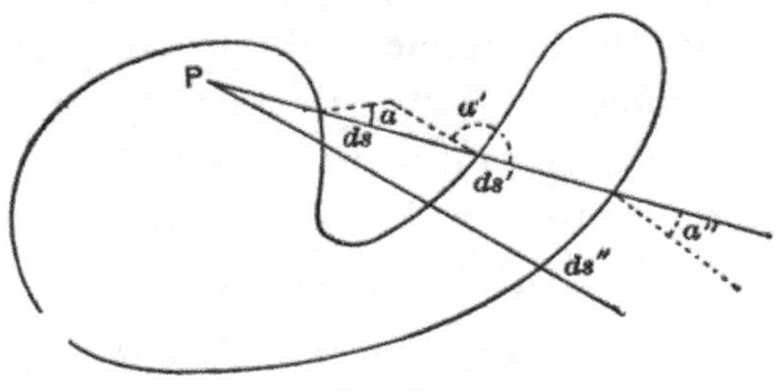

FIG. 4.

a point P, within a surface which, to make our treatment more general, has a re-entrant portion (Fig. 4).

From the point P as apex draw the elements of a cone corresponding to a solid angle $d\omega$, which is measured by the surface intercepted by the cone on a sphere of centre P and radius equal to unity.

Call ds, ds', ds'' the areas bounded by the intersections of the cone with the surface described; $d\omega$ represents the apparent surface of these intersections as seen from the point P. Call l, l', l'' the distances of the intersected elements from the point P; and $\alpha, \alpha', \alpha''$ the angles of the axis of the cone with the normals to the elements.

The flux of force traversing these elements are respectively

$$+ \frac{km}{l^2} \cos \alpha ds, \quad + \frac{km}{l'^2} \cos \alpha' ds', \quad + \frac{km}{l''^2} \cos \alpha ds''.$$

But

$$\frac{ds \cos \alpha}{l^2} = - \frac{ds' \cos \alpha'}{l'^2} = \frac{ds'' \cos \alpha''}{l''^2},$$

since these expressions, by definition, measure the solid angle $d\omega$. The flux reduces then to $km d\omega$, whatever be the number of intersections, provided that the number is uneven.

The total flux across the surface is given by the sum of all the elementary cones that can be drawn about P; thus

$$\int_0^{4\pi} km d\omega = 4\pi km.$$

4π represents the total surface cut by the cones on a sphere of unit radius.

II. If we had considered a mass m, outside of the closed surface, the elementary cones can traverse this surface only an even number of times, and would thus give a zero resultant.

III. Finally, if we suppose in the field masses, $m_1, m_2, m_3,$ some of which are in the interior and others on the exterior of the surface, the total flux through the surface will be the sum of the flux due to the masses within, Σm :

$$\int H ds \cos \alpha = 4\pi k \Sigma m.$$

21. Corollary I.—Suppose that the closed surface be bounded by the lateral walls of a tube of force and by two sections of this tube, s and s'.

As the walls of the tube cut no lines of force, the flux of force traversing the closed surface is limited to the flux $\int H \mathrm{d}s - \int H' \mathrm{d}s'$, $\mathrm{d}s$ and $\mathrm{d}s'$ being normal equipotential sections of the tube.   Consequently,

$$\int H \mathrm{d}s - \int H' \mathrm{d}s' = 4\pi k \Sigma m.$$

If there are no masses within this closed region,

$$\int H \mathrm{d}s = \int H' \mathrm{d}s'.$$

The flux entering by one base issues from the other, that is to say that the *flux is constant in a tube of force, as long as the tube does not encounter acting masses.*

This property, comparable with that of fluid circuits in which the flow remains constant as long as no outflowing sources are met, justifies the name *flux*, given to the mathematical expression which we have been considering.   We shall see that this property, known by the name of *continuity of flux*, plays an important part in electric and magnetic phenomena.

22. Corollary II.—If the tube of force were infinitely thin, we should have

$$Hds = H'ds' = dN;$$

whence

$$\frac{H'}{H} = \frac{\mathrm{d}s}{\mathrm{d}s'}.$$

In such a tube the intensity of the field is in inverse ratio to the section normal to the axis.   In a uniform field the tubes of force are necessarily cylindrical.

23. Corollary III.—The expression $H = \dfrac{\mathrm{d}N}{\mathrm{d}s}$ shows that the *intensity of a field is the flux per unit equipotential surface at the point under consideration.*

24. Unit Tube. Number of Lines of Force.—A tube chosen so that the expression $\int H ds = 1$ is assumed a *unit* tube.

Following a convention due to Faraday and admitted by many authors, the number of lines of force of a field, which in reality is indefinite, is limited to the number of unit tubes of which they form the axes.

In accordance with this convention, Gauss' theorem is enunciated as follows: The number of unit tubes or lines of force traversing a closed surface in a field is equal to $4\pi k$ times the sum of the quantities of the agent enveloped by this surface.

25. Potential Energy of Masses Subjected to Newtonian Forces.—In consequence of the repulsion exerted between masses of the same kind, a certain amount of work must be done to bring the masses m, m', m'' to the neighboring points o, o', o''. This work is stored up in the system in the state of potential energy, and is restored when the masses, being set free, separate indefinitely from one another under the effect of their mutual actions.

To determine the expression for the work done, suppose that the masses are formed by means of elementary masses brought up successively to the given points o, o', oo''.

To bring an element dm to a point o whose potential is U, we must by definition do an amount of work equal to Udm.

For the other points we obtain in the same way the elements of work $U'dm'$, $U''dm''$.

But in proportion as the masses increase the potential of each of the points rises. At the point o, for example, it passes from the value zero to the value

$$U = k\Sigma\frac{m}{r}.$$

We may suppose that the progressive increase of the masses takes place in a constant ratio, so that at a given instant the masses accumulated at the various points reach the same fractional part of their full values. In these conditions the potentials increase in the same ratio; the mean value of the potential at the point o is $\dfrac{U}{2}$, corresponding to elementary work, $\dfrac{U}{2}\, dm$.

The work necessary for the formation of the mass m is

$$w = \frac{U}{2} \int_{0}^{m} dm = \frac{Um}{2}.$$

The sum of the work expended on the various masses of the system will be

$$W = \frac{1}{2} \Sigma Um.$$

Hence we see that *the potential energy of the system is the half sum of the products of the masses by their potentials.*

The above assumption of proportional increments of the masses does not in the least invalidate the generality of the conclusion. For the potential energy is measured by the work stored, which depends only on the final state of the system, and is in no way dependent on the method by which the masses have come to this state. We shall also give a demonstration quite independent of such hypothesis.

When two masses m, m', at a distance l, separate still further by a length dl, the increase of potential energy is equal and of opposite sign to the work accomplished. We have therefore

$$dw = -\, k\frac{mm'}{l^{2}} dl.$$

When the masses move to an infinite distance from each other, the work done represents their total initial energy, that is

$$w = k\frac{mm'}{l}.$$

For a system of masses we get an expression of the form

$$W = k\Sigma\frac{mm'}{l}.$$

Observe that

$$\Sigma\frac{mm'}{l} = \frac{1}{2}\Sigma m\Sigma\frac{m'}{l},$$

the factor $\frac{1}{2}$ being necessary in order to avoid taking each couple of quantities twice.

But $k\Sigma\frac{m'}{l}$ represents the potential of the point at which the mass m is situated. We get, therefore,

$$W = \frac{1}{2}\Sigma m U.$$

APPLICATIONS.

Before going further let us apply the properties just

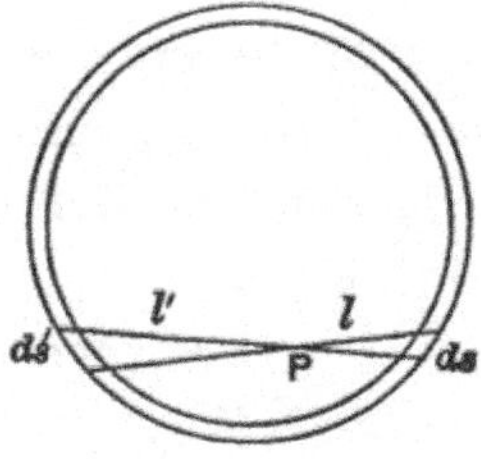

FIG. 5.

demonstrated to some simple cases which we shall later on meet with in certain electric and magnetic combinations.

26. I. An infinitely thin homogeneous spherical shell

exercises no action upon a mass within it.—A mass equal to unity being concentrated at P, draw from this point as apex an elementary cone cutting two surfaces ds and ds' on the sphere at distances l and l'. Let σ be the surface density, or quantity of the agent per unit of surface. The elements ds and ds' are consequently charged with masses

$$dm = \sigma ds,$$

$$dm' = \sigma ds'.$$

Their actions on the unit mass at P are

$$\frac{k\,dm}{l^2} = \frac{k\sigma ds}{l^2},$$

$$\frac{k\,dm'}{l'^2} = \frac{k\sigma ds'}{l'^2}.$$

But the elements ds and ds', making equal angles with the axis of the cone, are to each other as their projections perpendicular to this axis, and as these latter are themselves proportional to the square of their distances from the apex of the cone, we have

$$\frac{ds}{l^2} = \frac{ds'}{l'^2};$$

whence

$$\frac{k\,dm}{l^2} = \frac{k\,dm'}{l'^2}.$$

As the whole surface of the sphere can be divided into pairs of elements like ds and ds', whose actions neutralize each other, the total effect of the shell on the point P, or on any other internal point, is zero.

We conclude from this that *the potential is constant in all points inside of a spherical shell.*

This potential is therefore the same as that of the centre, which is expressed by

$$U = k\,\Sigma\,\frac{m}{L} = k\,\frac{M}{L} = 4\pi k L \sigma.$$

Corollaries.—(*a*) The surface of the shell under consideration is equipotential.

(*b*) The potential energy of the shell is $\dfrac{1}{2}k\dfrac{M'}{L}$.

(*c*) This conclusion would still hold in the case of a series of concentric shells acting upon an internal point, which may, at the limit, be situated on the internal surface of the innermost shell.

27. II. The action of a homogeneous spherical shell on an external point is the same as if the whole mass

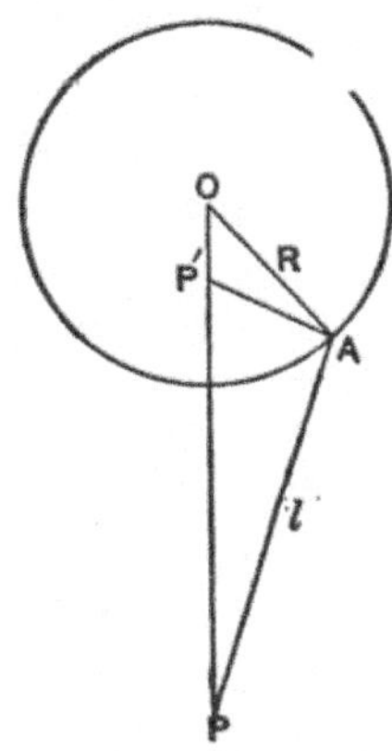

Fig. 6.

were concentrated at the centre of the sphere.—If unit mass be concentrated at P, it is evident that by symmetry the resultant action must be directed along OP. An element ds at A exerts on unit mass a force whose component along OP is

$$dH = k\sigma\frac{ds}{l^2}\cos\alpha.$$

P' being the conjugate point of P, such that $OP' \times OP = R^2$, connect A and P'. The triangles OAP' and OAP

are similar, for they have the angle AOP in common and the adjacent sides proportional; whence

$$\frac{l}{\overline{AP'}} = \frac{OP}{R};$$

and, substituting,

$$dH = k\sigma \frac{ds \cos \alpha}{\overline{AP'^2}} \times \frac{R^2}{\overline{OP^2}}.$$

But

$$\frac{ds \cos \alpha}{\overline{AP'^2}} = d\omega,$$

ω being the solid angle subtended by the element ds at the point P'; therefore

$$dH = k\sigma d\omega \times \frac{R}{\overline{OP^2}}.$$

The action of the whole shell will be

$$H = k\sigma \times \frac{R^2}{\overline{OP^2}} \int_0^{4\pi} d\omega = \frac{4\pi k R^2 \sigma}{\overline{OP^2}} = \frac{kM}{\overline{OP}}.$$

This action is the same as if the entire mass were concentrated in the point O.

Corollary.—For a point infinitely near to the surface the action of the shell would be $4\pi k\sigma$.

28. Action of a Homogeneous Sphere upon an External Point.—If the sphere were composed of a number of similar shells superposed, this conclusion would still hold. We can therefore say that *a homogeneous sphere, or sphere composed of homogeneous shells, acts upon an external point as if the mass were concentrated at the centre of the sphere.*

Calling in this case δ the mass per unit volume, we have

$$H = \frac{4}{3}\pi k R^3 \delta \times \frac{1}{\overline{OP^2}}.$$

This property justifies the hypothesis of the concentration in physical points of masses which in reality occupy definite volumes around these points.

As a particular case, if the point is at the surface of the sphere, the preceding expression reduces to

$$H = \frac{4}{3}\pi k R \delta.$$

29. Action of a Homogeneous Sphere upon an Internal Point.—If the point were inside of a homogeneous sphere, this latter could be divided into two parts, separated by a concentric sphere passing through the given point.

The action of the external portion is null; the action of the sphere internal to the point is equal to that of an equal mass concentrated at the centre. Denoting by l the distance of the point P from the centre,

$$H = \frac{4}{3}\pi k l \delta.$$

30. Surface Pressure.—In the case of a homogeneous

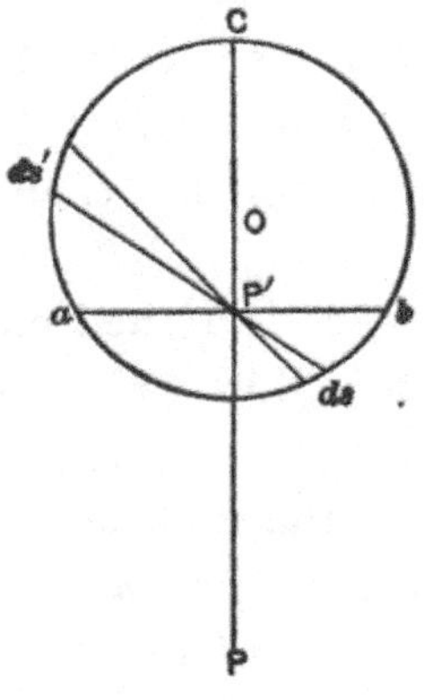

Fig. 7.

spherical shell the component due to the element d*s* depends only on the solid angle subtended by it at the point

P'. It is therefore equal to that of the element ds', corresponding to ds.

The same holds for all the elements of the segment *adb*, taken in pairs with those of the segment *acb*. The plane projected on *ab* divides the sphere into two zones exercising the same actions upon P, equal to

$$\frac{2\pi k R^{2} \sigma}{\overline{OP}^{2}}.$$

If the point P is removed indefinitely from the sphere, the two segments tend to become equal. If, on the other hand, the point P approaches indefinitely to the spherical shell, one of the segments has as its limit the entire sphere, while the other tends towards zero.

In this last case the preceding expression becomes $2\pi k\sigma$.

Now we have just seen that the whole shell exerts upon unit mass, situated infinitely near to its surface, an action equal to $4\pi k\sigma$. It follows from this that the infinitely small element adjoining the point exercises an action equal to that of the entire sphere; and in fact when the point P traverses the shell the force acting upon it becomes zero; during this infinitesimal displacement the action of the spherical shell has remained constant, but that of the surface element next to the point has changed sign, so that the resultant is zero.

If the action of the spherical shell on unit mass situated at its surface is $2\pi k\sigma$, it will be $2\pi k\sigma$, on the mass σ which charges unit surface.

This force, with which the shell acts upon the charge of unit surface, is called *surface pressure*.

The intensity of the field infinitely near the surface being $H = 4\pi k\sigma$, the surface tension is expressed indifferently by

$$2\pi k\sigma, \quad \text{or by} \quad \frac{H^{2}}{8\pi k}.$$

31. Potential Due to an Infinitely Thin Disk Uniformly Charged.

—Let σ be the surface density of a disk projected

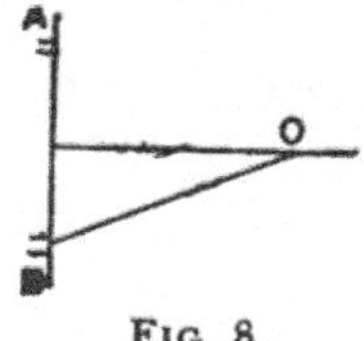

FIG. 8.

on AB (Fig. 8). On a ring concentric with the disk, with radius r and thickness dr, the charge is $2\pi r dr \sigma$.

The potential due to this ring at a point O on the axis of the disk and at a distance a is

$$dU = \frac{2\pi k r dr \sigma}{\sqrt{r^2 + a^2}}.$$

The potential due to the whole disk, whose radius is R, will be

$$U = \int_0^R \frac{2\pi k r dr \sigma}{\sqrt{r^2 + a^2}} = 2\pi k \sigma (\sqrt{R^2 + a^2} - a).$$

The intensity of the field at the point O is

$$H = -\frac{dU}{da} = -2\pi k \sigma \left(\frac{a}{\sqrt{R^2 + a^2}} - 1 \right) = 2\pi k \sigma (1 - \cos \alpha),$$

α being the plane angle subtended by the radius of the disk at the given point.

At a point infinitely near the surface of the disk we have

$$H = 2\pi k \sigma.$$

This expression represents the force due to the charge of the element infinitely near to the point (§ 30). The other elements exert forces which mutually neutralize each other.

It will be noticed that $2\pi(1 - \cos \alpha)$ expresses the angle subtended by the disk at the point O. We obtain directly the expression for the intensity of the field by considering

the action of an element d*s* of the disk upon unit mass.
Calling α the angle of the axis with the right line which
joins the element to the point O, the projection of the force
due to this element along the direction of the axis is

$$dH = k\sigma ds \frac{1}{l^2} \cos \alpha = k\sigma d\omega,$$

$d\omega$ expressing the solid angle subtended by the element at
the point O.

The action of the whole disk is $H = k\sigma \omega$.

When the unit mass is infinitely near the disk, the force
is $2\pi k\sigma$. Moreover, it is the same at all points of the disk.

MAGNETISM.

PROPERTIES OF MAGNETS.

32. Definitions.—The name *magnet* is given to those bodies which possess the property of attracting iron filings. The lodestone or magnetic oxide of iron possesses this property by nature, but it is artificially acquired in a much higher degree by iron and its derivatives, steel and cast iron. Tempered steel is the substance which retains in the greatest degree the attractive power developed by *magnetization*.

When a magnetized steel bar is thrust into iron filings, it is noticed that the filings cling by preference to certain parts of the bar, designated by the name of *poles*. These bars generally present two poles separated by a *neutral region* having a feeble or no action upon iron filings.

33. Action of the Earth on a Magnet.—When a magnet is suspended by its centre of gravity, one of its poles is invariably directed towards the north, the other towards the south. For this reason the first pole is called the *north* or N-*pole*, the second, the *south* or S-*pole*.

34. Law of Magnetic Attractions.—If several magnets are placed near each other, it is observed that the poles of the same name repel each other and that those of contrary name attract each other. The study of these actions presents some difficulties, because it is not possible to investigate the reciprocal action of two isolated poles.

Coulomb, however, observed that long magnetized rods have their centres of action near their extremities. By bringing the poles of two of these rods sufficiently near together he was able to almost entirely eliminate the action of the opposite poles.　He discovered experimentally that *magnetic forces decrease in inverse ratio to the square of the distance between the acting poles.*

We shall see further on that this law has been rigorously verified by Gauss.

Magnetic actions are therefore to be classed among the Newtonian forces, and we may apply to them the general theorems demonstrated in the introduction to this work, the agent acting in this case being magnetism.

We shall designate by *quantity of magnetism* or *mass of a pole* a quantity proportional to the force which it exerts upon neighboring poles.

Let m and m' be the quantities of magnetism of two poles: their mutual reaction is by definition proportional to m and m', and consequently to the product mm'.

The expression of the force is therefore

$$f = k\frac{mm'}{l^2},$$

l expressing the distance between the two poles.

35. Unit Pole.—The coefficient k in the preceding expression may be considered arbitrary and taken equal to unity.　In reality, the coefficient k varies with the medium in which the magnetic bodies are situated, but the differences in the various gaseous media in this respect are very small at ordinary temperatures.　Consequently the unit quantity of magnetism is the quantity which repels an equal quantity, at a distance of one centimetre, with the force of one dyne; its dimensions are

$$[L^{\frac{3}{2}}M^{\frac{1}{2}}T^{-1}].$$

If the poles are of contrary name, the force becomes attractive.　To deduce this from the preceding formula, we need only give opposite signs to poles of contrary name.　It has been settled that the N pole shall have the $+$ sign, and the S pole the $-$ sign.

36. Definitions. — Applying the general definitions adopted in the Introduction, we will call the space in which magnetic forces exist a *magnetic field. The intensity at a point of the field* is measured by the action it exerted there on positive unit mass. The direction of this action gives the *direction of the field.* A *magnetic line of force* represents the path of an infinitesimal positive mass free to move in the field. The field possesses a potential called *magnetic potential*, whose expression for a point situated at distances l, l', l'' ... from masses m, m', m'' ... is

$$U = \Sigma \frac{m}{l},$$

each mass being given its own sign.

By this definition the dimensions of magnetic potential are

$$[L^{\frac{3}{2}} M^{\frac{1}{2}} T^{-1}].$$

The component of field intensity in a direction l, at a point where the potential is U, is expressed by

$$\mathcal{H} = - \frac{dU}{dl},$$

dimensions,

$$[L^{-\frac{1}{2}} M^{\frac{1}{2}} T^{-1}].$$

This expression represents, as we have seen in §23, the flux of force per unit of surface normal to dl.

The dimensions of field intensity differ from those of a force. It is the force per unit pole, and its dimensions are those of a force divided by those of a magnetic mass.

The most intense fields that have been produced up to the present reach about 30,000 C. G. S. units, or 30 kilogausses; such a field develops a force of 30,000 dynes, or about 30 grams, upon unit pole.

37. Action of a Uniform Field on a Magnet.—Consider a uniform field, in which the intensity $\mathcal{K}$ is constant in magnitude and direction. Experiment shows that in such a field a magnet is not subjected to any force of translation, but that it simply tends to place itself in a definite direction. Hence we conclude that the sum of the positive masses of the bar is equal to the sum of the negative masses, since the resultant of the actions of the field on the first, or $\mathcal{K}\Sigma m$, is balanced by the resultant $\mathcal{K}\Sigma(-m)$ of the actions on the second.

These resultants are applied at two points which, in the mathematical theory of magnetism, receive more particularly the name of *poles*. It must be observed, however, that the poles which are thus defined have no more physical existence than has the centre of gravity of a body. An imaginary line passing through the poles is called the *magnetic axis* of the magnet. The distance l between the poles is the true *length* of the magnet.

The product of the magnetic mass at one pole by the distance between the poles is the *magnetic moment* of the bar $(\mathcal{M})$:

$$l\,\Sigma\,m = \mathcal{M}.$$

Designating by β the angle of the magnetic axis with the direction of the field, the couple acting on the magnet is expressed by

$$\mathcal{K}\,l\sin\beta\,\Sigma m = \mathcal{K}\,\mathcal{M}\sin\beta.$$

If the magnet be suspended by its centre of gravity, the duration of a complete oscillation of small amplitude is

$$t = 2\pi\sqrt{\frac{K^2}{w}},$$

where K^2 is the moment of inertia of the magnet, and w the maximum couple, equal to $\mathfrak{KM}$.

38. Terrestrial Magnetic Field.— Experiment shows that the duration of the oscillation of a magnet is constant within the limits of a laboratory room, provided that there be no other magnetic mass in the room or near by. It may therefore be assumed that, in a space of small extent, the earth develops a uniform field. The vertical plane passing through the axis of a magnet hanging freely is called the *magnetic meridian* of a place.

The *declination* is the angle of this plane with the geographical meridian; the *inclination*, the angle of the axis of the magnet with the horizontal. In our hemisphere the north pole of magnets dips below the horizon, so that a weight must be placed on the south pole to make the oscillations take place in the horizontal plane. The component of the intensity of the terrestrial field in this plane is called the *horizontal component*.

The terrestrial lines of force, which, in a limited space, may be considered as parallel, really converge towards points called *magnetic poles*, which oscillate in the neighborhood of the geographical poles.

The following numerical data, due to Airy, show the mean magnetic values for Greenwich, t being the date :

Declination : $19° \ 12.1' — (t — 1876) \times 7.38'$.

Horizontal component : $0.1797 + (t — 1876) \times 0.00027$.

Inclination : $67° \ 40.3' — (t — 1876) \times 2.04.'$[*]

We see from this that the terrestrial magnetic field has only a feeble intensity. It will be shown later on that it is possible to produce very much intenser fields in the interior

---

[*] For the study of terrestrial magnetism consult Gauss, *Allgemeine Theorie des Erdmagnetismus;* Mascart et Joubert, *Leçons sur l'Électricité et sur le Magnétisme*, Vol. I.

of a bobbin of wire traversed by an electric current. In such a bobbin, if the length is very great in proportion to the cross-section, the field may be considered uniform.

39. Weber's Hypothesis.—When a magnet is broken across its neutral region, we do not get two isolated poles, but two new magnets. The original magnet can be restored by joining the broken pieces together again; the poles which were developed at the surfaces of rupture neutralize each other. This fact of the division of a magnet always furnishing complete magnets, however small the fragments, may be, has led Weber to suppose that the polarization takes places in the molecules composing the bar, each one of them being a complete magnet possessing two poles.* The neutral state, by this hypothesis, results from the non-orientation of the molecules, whose poles mutually neutralize each other. But if a neutral bar is placed in a field, the magnetic axes of the molecules are drawn into their proper positions: the N-poles in the direction of the field, and the S-poles the opposite way. To explain the observed variations in the degree of magnetization, we have to assume that the molecules oppose a certain resistance to this alignment, varying according to the physical condition of the bar, and to which the name *coercive force* has been given.

This resistance, an explanation of which by Ewing will be seen latter on, is feeble in annealed or soft iron, so that when a bar of this metal is introduced into a magnetic field of medium intensity it becomes strongly magnetized. But it readily loses its magnetization when removed from the field and retains only traces of *residual* magnetism. The name *magnetizing force* is given to the intensity of the field which induces the magnetization.

---

* See Maxwell, *Electricity and Magnetism*, Vol. II.

Cold hammering increases the coercive force of iron, but this force is especially increased by combining the metal with certain foreign substances, such as carbon, tungsten and chromium, in small proportions. A steel bar attains its maximum coercive force when it has been heated to a bright red, and *tempered* either by sudden chilling in oil, water, or mercury, or by powerful pressure under an hydraulic press during its cooling. This last method gives the metal a more uniform hardness than tempering by immersion. The degree of annealing may be estimated by the electric conductivity of samples. This conductivity increases from its original value for soft steel to three times that value for steel tempered in mercury. A tempered steel bar is more difficult to magnetize than iron, but it retains a considerable permanent magnetization.

Different facts tend to corroborate Weber's hypothesis.

1. The magnetization in a field which is increasing in intensity tends towards a limit called *saturation*, which probably corresponds to the parallelism of the molecular axes with the direction of the field.

2. Every cause of molecular disturbance favors the magnetization of a bar subjected to a magnetizing force, and also favors its demagnetization after it has been withdrawn from the field. Thus a bar of soft iron placed vertically is magnetized by the action of the vertical component of terrestrial magnetism when it is struck lightly.

The bar retains its magnetization when placed in a different position, provided it be kept free from vibrations, but a slight blow is sufficient to dissipate its magnetism. Vibrations have a specially marked effect upon iron. Ewing has shown that if a bar of this metal be kept from the slightest vibration one can obtain residual magnetizations much greater than those shown in steel bars; but the least vibration causes the acquired magnetism to vanish almost completely. A violent blow can likewise take from a recently magnetized bar of steel more than half its magnetic moment, but successive blows produce a more and more feeble effect.

The same holds for variations of temperature. Raising the temperature weakens the power of a magnet. At a bright red heat the magnetization disappears entirely.

If a magnetized and tempered bar is annealed at a temperature of 100°, for example, constancy of the magnetization for lower temperatures is secured; that is to say, variations less than 100° cause only a temporary change in the magnetic moment of the bar, which resumes the same value at the same temperature.

The variations of the magnetic moment are then sensibly a linear function of the temperature; let $\mathfrak{M}_0$ be the moment at 0° C., $\mathfrak{M}_t$ the moment at $\theta°$ C.:

$$\mathfrak{M}_t = \mathfrak{M}_0(1 - \alpha\theta).$$

3. Magnetization produces a slight progressive elongation of the magnetized bars up to a certain limit, beyond which a contraction takes place, as if the molecules, after being oriented, tended to approach each other and to diminish the intermolecular space (Bidwell).

When a bar is subjected to variable magnetizing forces, it produces a sound, which may be attributed to the displacements of air-molecules caused by the dilatations or contractions of the magnet.

4. Beetz has shown that a *feeble* magnetizing force applied to iron at the moment of its precipitation by electrolysis magnetizes it to saturation, which is the result of the fact that the molecules of the metal change their position freely at the moment of reduction.

40. Elementary Magnets. Intensity of Magnetization. —The hypothesis just developed leads us to analyze the properties of *elementary magnets*, whose length is taken as infinitely small in comparison with the finite distances of the field.

Intensity of magnetization of an elementary magnet is the name given to the ratio of its magnetic moment to its volume.

$$\mathfrak{I} = \frac{\mathfrak{M}}{V}; \quad \text{dimensions}, \quad [L^{-\frac{1}{2}}M^{\frac{1}{2}}T^{-1}].$$

Elementary magnets are, by assumption, cylindrical in form, with their poles concentrated on the two ends. Under these conditions we call the ratio of the magnetic mass of the poles to their surface the *density* of the poles:

$$\sigma = \frac{m}{s}.$$

Now, calling the length of the magnet l, we have $ml = \mathfrak{M}$, $sl = V$; it follows that the density and intensity of magnetization have the same numerical expression and the same dimensions.

Let SN (Fig. 9) be an elementary magnet whose poles of

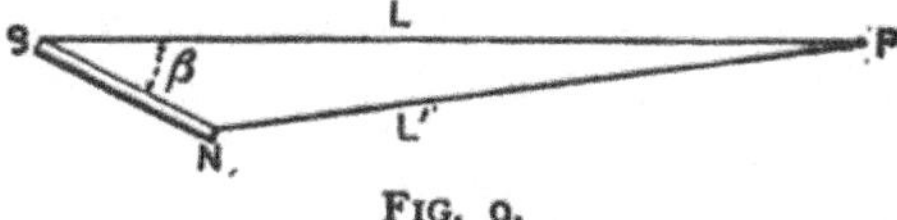

FIG. 9.

mass m are at distances L, L' from a point P. The magnetic potential in this point is

$$U = \Sigma \frac{m}{L} = m\left(\frac{1}{L'} - \frac{1}{L}\right) = m\frac{L-L'}{LL'}.$$

As L differs from L' by only an infinitely small quantity, we can replace LL' by L^2, and $L - L'$ by $l \cos \beta$, β being the angle made by the axis of the magnet SN with the right line drawn from S to P. Consequently

$$U = \frac{ml \cos \theta}{L^2} = \frac{\mathfrak{M} \cos \theta}{L^2}.$$

41. Magnetic or Solenoidal Filament.—If we place elementary magnets of equal intensity of magnetization end to end, the adjoining poles neutralize each other, and there remain only two resultant poles at the ends of the chain, which is called a *magnetic filament* or *solenoidal filament*.

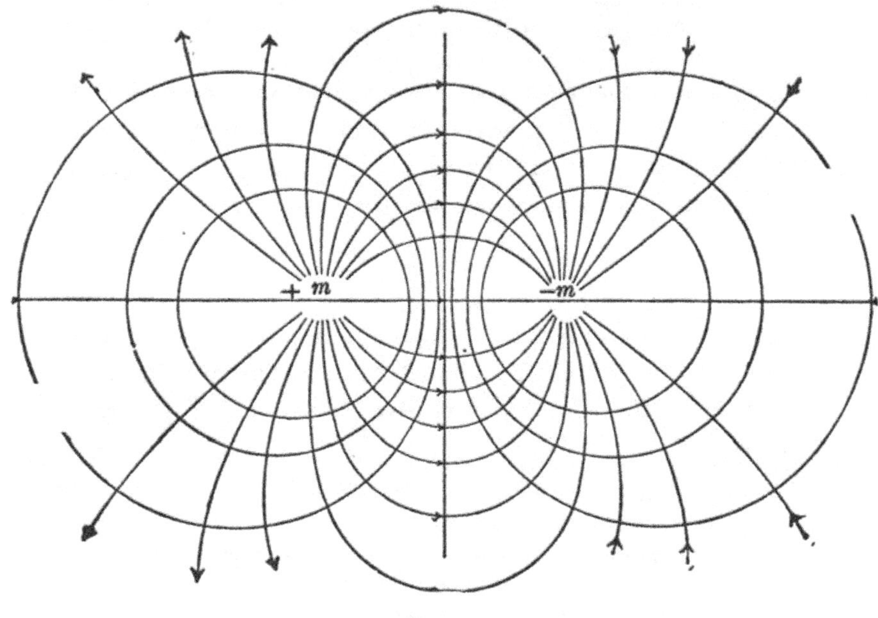

FIG. 10.

The potential at a point P, situated at distances L and L' from the resultant poles, is

$$U = m\left(\frac{1}{L'} - \frac{1}{L}\right),$$

which expression is independent of the form of the magnetic filament, but depends solely on the position of its extremities. Consequently a magnetic filament forming a closed chain has a zero potential for all external points; that is, a closed magnetic filament exercises no action upon neighboring magnetic masses.

The equation

$$m\left(\frac{1}{L'} - \frac{1}{L}\right) = \text{constant}$$

represents an equipotential surface.

Adopting the graphic method shown in § 16, we can draw the equipotential lines in a plane passing through the masses $+ m$ and $- m$.

Fig. 10 shows the equipotential lines due to two such masses; they are ovoid curves enveloping the poles. The vertical axis of symmetry is equipotential. The lines of force, whose direction is shown by the arrows, cut the equipotential lines orthogonally.

42. Uniform Magnets.—Imagine a right circular cylinder formed of equal adjoining rectilinear filaments. The effect of such a combination, as regards external points, is the same as that of two magnetic shells, of equal and opposite densities, situated at the extremities of the cylinder. The intensity of magnetization is constant at every point of the cylinder.

The action of a right cylinder, uniformly magnetized, at an exterior point on the prolongation of its axis is, denoting by α and α' the plane angles subtended by the radii of the bases of the cylinder at the given point (§ 31),

$$\mathcal{K} = 2\pi\mathfrak{J}(\cos \alpha' - \cos \alpha).$$

We can likewise place together rectilinear filaments of equal intensity of magnetization, but of different lengths, whose extremities abut on the surface of a sphere. A spherical magnet thus defined presents two hemispherical shells of opposite signs.

The density of these shells is equal to the intensity of the filaments at the extremities of the diameter NS, where the elements cut the surface of the sphere normally. At other points the density decreases as the cosine of the angle α, α representing the inclination of the filaments to the radii of the sphere. It is zero along the great circle normal to *NS*.

Such a distribution may be represented by slipping a positive sphere, with centre O, over an equal sphere, but of contrary sign, with centre at O', the distance between the centres being infinitesimal, so that OO' represents the maximum thickness of the magnetic shell.

In the meridian section represented in Fig. 11, the shells bounded by the surfaces of separation of the two spheres

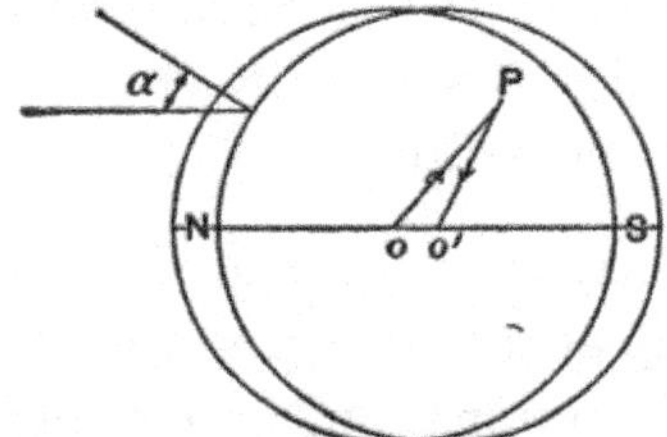

FIG. 11.

have a thickness decreasing as the cosine of the angle α. The region common to the two is evidently neutral.

The effect of the system on an internal point P is equal to the resultant of the actions of both spheres.

Now calling δ the cubic density of the masses, the effect of the sphere O on unit pole situated at P is (§ 29)

$$\frac{4}{3}\pi\delta \times OP.$$

This action, directed along OP, may be represented by the length of this line. The negative sphere exercises an action equal to

$$-\frac{4}{3}\pi\delta \times O'P,$$

and represented by PO'.

The resultant of these two forces is

$$\frac{4}{3}\pi\delta \times OO',$$

since OO' closes the triangle formed by the components OP and PO'. Now $\delta \times OO' = \sigma$ represents the maximum magnetic density of the shell.

The resultant of the actions of a uniformly magnetized sphere is therefore constant in magnitude and direction for all internal points, and equal to $\frac{4}{3}\pi$ multiplied by the maximum density or intensity of magnetization of the sphere.

43. Magnetic Shells.—This name is given to a lamellar magnet whose faces have equal and opposite densities. A magnetic shell may be considered as the result of the juxtaposition of elementary magnets.

Let σ be the density of the faces, equal to the intensity of magnetization of the component elements; ϵ the thickness of the shell, that is, the length of the elements.

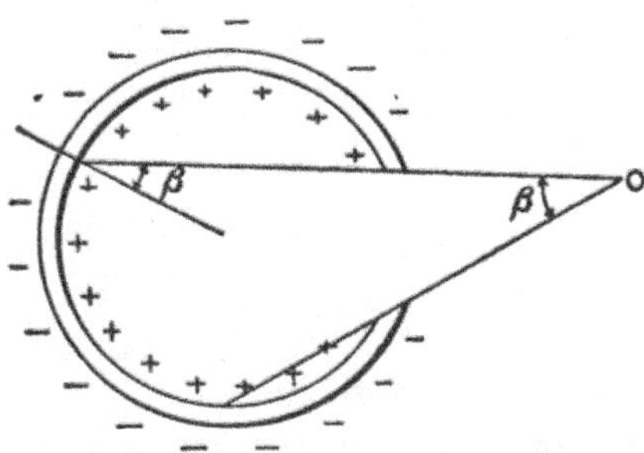

Fig. 12.

The product $\epsilon\sigma = \mathfrak{F}_{,}$, which represents the moment of the shell per unit surface, is called the *strength* of the shell. The dimensions of this quantity, the importance of which will be understood in the study of electromagnetism, are

$$[L^{\frac{1}{2}}M^{\frac{1}{2}}T^{-1}].$$

Let C (Fig. 12) be a curved shell with its positive face

turned towards the interior. The potential at a point O is the sum of the potentials due to the component elementary magnets. For an element ds, whose axis makes an angle β with the right line joining it to the point O, situated at a distance L, the potential is

$$dU = \frac{\epsilon\sigma ds \cos \beta}{L^2} \, ;$$

but $\dfrac{ds \cos \theta}{L^2}$ represents the solid angle subtended by the surface ds at the point O. In order to make this notation uniform with that which will be met with in electromagnetism, we shall consider the solid angle as positive when the point O faces the S pole of the magnetic element.

We have then

$$dU = - \mathfrak{F}_{,}d\beta.$$

Extending the integration to all the elements of the shell, and observing that all the elementary cones which cut the

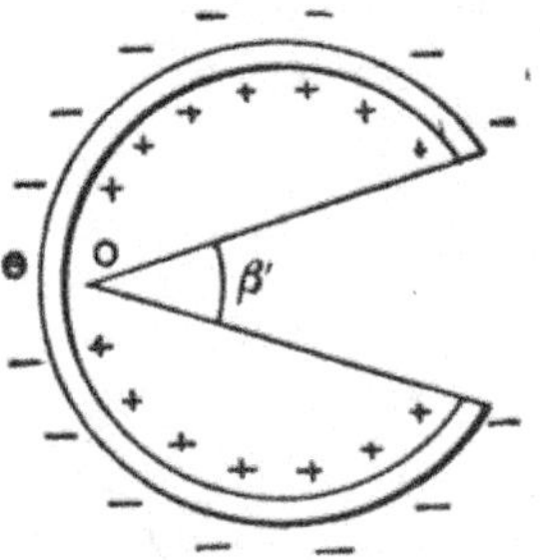

FIG. 13.

shell twice give equal and contrary elements of potential, we get for the total potential

$$U = - \mathfrak{F}_{,}\beta.$$

The potential due to a shell at any point is equal to the product of the strength of the shell by the solid angle subtended at that point by the contour of the shell.

44. Corollary.—If the point O were at the interior of the shell, the solid angle would be $+ (4\pi - \beta')$. Consequently, if the point passes from a point O to a point O' infinitely near on the other side of the shell, the potential varies from $+ \mathfrak{F}_s(4\pi - \beta')$ to $- \mathfrak{F}_s\beta'$, that is, by the quantity $4\pi\mathfrak{F}_s$.

The work done by unit positive mass, in passing from a point on the surface of a shell to a point infinitely near situated on the other side, is equal to 4π multiplied by the strength of the shell. This work is, moreover, independent of the path followed between these two points (§ 12).

45. Energy of a Shell in a Field.—Let us consider a field of force due to a pole m situated at a point O (Fig. 12); the work expended to bring the shell to its present position represents the relative energy of the shell and the field. It is equal to the work required to bring the mass m to the point O whose potential is U, or

$$mU = - m\mathfrak{F}_s\beta.$$

Now $m\beta$ is the flux of force from the pole across the solid angle limited by the contour of the shell.

We will call this flux Φ, considering it as positive when it enters by the negative face of the shell, and as negative when it enters by the positive face:

$$W = - \mathfrak{F}_s\Phi.$$

If the field is produced by several poles m, m', m'', the total energy will become

$$W = - \mathfrak{F}_s\Sigma m\beta = - \mathfrak{F}_s\Phi.$$

The relative energy of a shell and a field is therefore equal to the product of the strength of the shell by the flux included within the contour of the shell.

If the flux penetrates, as in Fig. 12, by the positive face, Φ is negative, and the product takes the plus sign.

When a shell is free to move in a field, it tends to move so that the expression for the potential energy becomes a minimum;

that is, the flux entering by the negative face tends towards a maximum. It will easily be shown that this condition is satisfied in the case of a shell and a positive pole when the latter touches the negative face. A plane shell situated in a uniform field will take up a position normal to the direction of the field, so that the lines of force penetrate it by the S-face.

46. Relative Energy of Two Shells.—Let us consider two neighboring shells A, A', of strengths $\mathcal{F}_s$ and $\mathcal{F}_s'$. Let Φ' be the flux of force from A' across the section of A entering by its negative face. The energy of the shell A is, as we have just seen, expressed by

$$W = -\,\mathcal{F}\Phi'.$$

Now the flux Φ' may be represented by the product of $\mathcal{F}'$ into a factor L_m; whence

$$W = -\,\mathcal{F}\mathcal{F}'L_m. \qquad \ldots \ldots \ldots \quad (1)$$

This expression must evidently represent the energy of the shell A', for the same work is expended to bring the shell A' up to A as to bring A up to A'. But as the energy of A' is also given by the product of $\mathcal{F}_s'$ into the flux Φ passing from A to A', we see that $\Phi = \mathcal{F}_s L_m$, just as we found $\Phi' = \mathcal{F}_s'L_m$; whence we obtain

$$L_m = \frac{\Phi}{\mathcal{F}_s} = \frac{\Phi'}{\mathcal{F}_s'}.$$

The factor L_m, called *coefficient of mutual induction* of the two shells, represents, as seen above, the ratio of the flux across one of the shells to the strength of the neighboring shell. Equation (1) shows that the dimensions of L_m reduce to $[L]$.

47. Artificial Magnets.—Instead of presenting, like uniform magnets, a surface distribution of magnetism, magnetized bars possess free magnetic masses internally. We can even superpose opposite magnetizations in a steel bar by submitting it successively to magnetizing forces of opposite directions. When such a bar is dissolved in an acid, there appear progressively layers magnetized in opposite directions.

This experiment proves that the magnetization affects at first the superficial layers of the bar, which are moreover tempered harder than the inner layers. Hence the utility of employing thin plates of steel, separately magnetized, in order to obtain powerful magnets.

We can show in a striking way the form of the magnetic

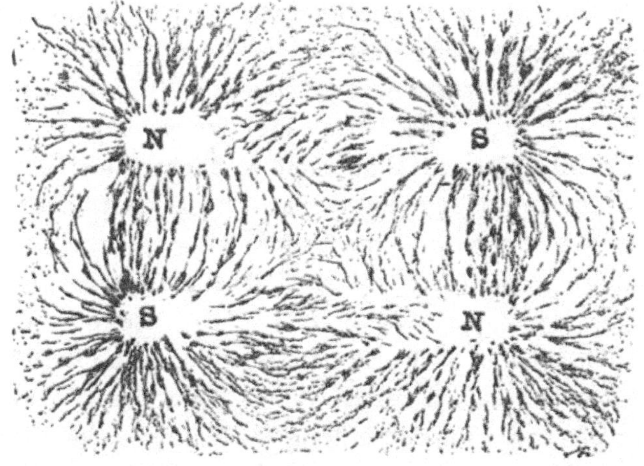

FIG. 14.

field due to a bar magnet by placing over it a sheet of paper covered with iron filings. The particles of iron become magnetized by induction, orient themselves along the directions of the field, and arrange themselves in continuous rows, representing the lines of force.

The image thus produced may be fixed if sensitized photographic paper be used and then exposed to the actinic action of light while covered with the filings. The development of the image shows the shadows produced by the filings. Fig. 14 thus represents the magnetic field of two adjacent bar magnets.

By observing the distribution of iron filings in the field of a single or of several magnets, and the curious patterns reproduced by the particles, Faraday was led to the idea that the seat of the magnetic forces is in the medium which separates the acting poles. According to him, lines of force are not a mere mathematical conception, but have a real existence corresponding to a particular state of the space around the poles. Faraday imagined this medium as being strained along the lines of force, and he readily substituted mentally for these lines elastic threads having a tendency to contract and thus cause neighboring poles to approach.

To explain the curvature of the lines of force, Faraday assumed that they repel each other when they proceed in the same direction, so that each of them takes a curved form whose tendency to return to a rectilinear form balances the repulsion of the neighboring lines.

Although experiment shows that a magnetized bar has free magnetic masses within it, it is possible to imagine a surface distribution of magnetism producing the same external field as the actual distribution. Suppose the magnet, for example, to be formed of longitudinal magnetic filaments some of which end on the end faces of the magnet and others on its lateral surfaces. Then the poles of the filaments constitute the surface charges of the magnet, as we have seen in the case of a magnetized sphere (§ 42), and the curves taken by the iron filings (Fig. 14), may be considered as the prolongation of the axes of the filaments.

To determine the imaginary distribution of magnetism giving the same external results as the real distribution, we measure the variation of the field intensity around the magnet along its axis. To do this, we determine the period of oscillation of a small magnetized needle moving on a pivot, and which is brought successively to the various points where it is desired to know the intensity. This means is not, however, rigorous, because the force is not the same at both poles of the needle, and the latter's magnetism may be altered under the influence of the field that is being investigated. By this experiment we find that the field decreases rapidly from the extremity of the magnet towards the middle, unless there be intermediate or *consequent* poles. In long needles of hard steel the poles are very near the ends, and the neutral line extends over the greatest part of their length. In every case the neighborhood of the edges gives a more intense field than the neighborhood of the plane surfaces.

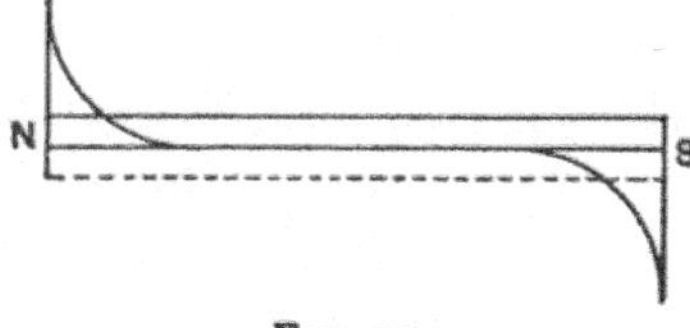

Fig. 15.

Fig. 15 shows the curves obtained by marking off on the perpendiculars to the axis of the bar lengths proportional to the components of the field along these lines. The ordinates are inversely proportional to the square of the periods of oscillation of the magnetized needle opposite various points of the axis and at the same distance from it, and oscillating in a plane normal to the magnet. These ordinates may be considered as proportional to the thickness of the magnetic shell having the same external effect as the real distribution of magnetism.

Magnets undergo a slow demagnetization, which may be explained by the repulsion exercised between poles of the same name in neighboring molecules. This loss is retarded by joining the poles by a piece of soft iron called *armature* or *keeper*. Opposite poles are developed in this latter which retain the magnetization of the magnet, since closed magnetic filaments are formed through the armature, and the poles of the elements composing these filaments attract and neutralize each other in couples.

The best steels for permanent magnets are those capable of acquiring the hardest temper.

The addition of 3 per cent. of tungsten increases the coercive force of steel very perceptibly. The tempering may be done in oil, water, or mercury. The bath should be of sufficient volume to prevent great rise in temperature and splashing of the liquid. According to Strouhal and Barus, the best way to obtain a powerful and constant magnet is to make the steel as hard as possible by tempering, and then to anneal it for 20 to 30 hours in steam at 100° C. It is next magnetized by placing its extremities on the poles of a powerful electromagnet, and finally annealing it again for at least 5 hours in steam. This method secures a magnetization which resists, as far as is possible, both blows and the daily variations of temperature.

According to Preece, the intensity of permanent magnetization obtainable in prisms of 1 cm section and 10 cm length, made of good magnet steel bearing the stamp Marchal, Clémandot & Allevard, varies from 100 to 225 C. G. S. units. These numbers express the ratio of the permanent moment of the magnets to their volume.

48. Determination of the Magnetic Moment of a Magnet. Magnetometer.—When we cause a magnet to oscillate horizontally in the earth's magnetic field, the magnet being hung by a thread having no sensible torsion-pull, the duration of a complete oscillation, of sufficiently small amplitude, is

$$t = 2\pi \sqrt{\frac{K}{\mathfrak{M}\,\mathfrak{H}}}$$

K being the moment of inertia of the magnet, $\mathfrak{M}$ its magnetic moment, and $\mathfrak{K}$ the horizontal component of the earth's field.

From this we deduce

$$\mathfrak{M}\mathfrak{K} = \frac{4\pi^{2}K}{t^{2}}. \qquad \ldots \ldots \quad (1)$$

If $\mathfrak{K}$ be known, the magnetic moment can be determined from this equation.

If this is not the case, a second experiment may be made. The magnet being placed in a horizontal plane and normal to the magnetic meridian, a small magnetized needle is hung

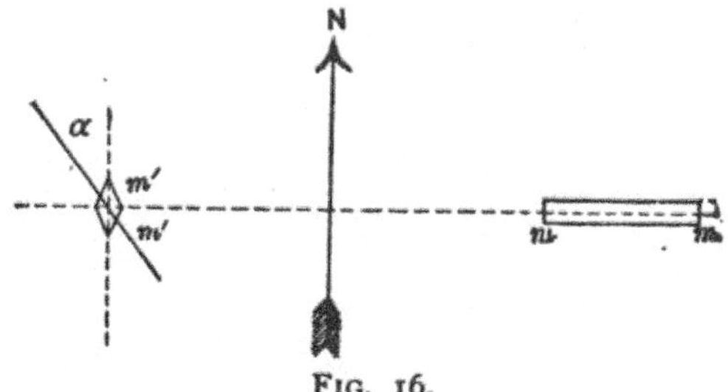

FIG. 16.

at a certain distance on the prolongation of the given magnet's axis (Fig. 16).

Let $2l$ be the distance between the poles of the magnet, $2l'$ that between the poles of the needle, L the distance from the centre of the magnet to that of the needle, $+m$ and $-m$ the poles of the magnet, $+m'$ and $-m'$ the poles of the needle.

The magnetic moments are respectively

$$\mathfrak{M} = 2ml,$$
$$\mathfrak{M}' = 2m'l'.$$

The movable needle, being drawn in one direction by the earth's magnetism and in the perpendicular direction by the magnet, takes up a position of equilibrium corresponding to an angle α between its axis and the magnetic meridian.

On account of the small dimensions of the needle, we may consider that the forces exerted upon its poles by the poles of the magnet are equal and opposite and form a couple.

The couple due to the earth's field is expressed by

$$w = \mathfrak{M}' \mathfrak{H} \sin \alpha.$$

The couple due to the magnet is, calling F the force which it exerts on the poles of the needle,

$$w' = 2Fl' \cos \alpha.$$

Now by Coulomb's law

$$F = \frac{mm'}{(L-l)^2} - \frac{mm'}{(L+l)^2} = mm' \times \frac{4Ll}{(L^2-l^2)^2};$$

consequently

$$w' = 2l'mm'\frac{4Ll \cos \alpha}{(L^2-l^2)^2} = 2\mathfrak{M}\mathfrak{M}'\frac{L \cos \alpha}{(L^2-l^2)^2}.$$

The condition of equilibrium

$$w = w',$$

gives

$$\mathfrak{M}'\mathfrak{H} \sin \alpha = 2\mathfrak{M}\mathfrak{M}'\frac{L \cos \alpha}{(L^2-l^2)^2};$$

whence

$$\frac{\mathfrak{M}}{\mathfrak{H}}\left(1 - \frac{l^2}{L^2}\right)^{-2} = \frac{L^2 \tan \alpha}{2}.$$

But

$$\left(1 - \frac{l^2}{L^2}\right)^{-2} = 1 + 2\frac{l^2}{L^2} + 3\frac{l^4}{L^4} + \cdots;$$

if l is sufficiently small, we may neglect the powers of $\frac{l}{L}$ higher than the second, and write

$$\frac{\mathfrak{M}}{\mathfrak{H}}\left(1 + 2\frac{l^2}{L^2}\right) = \frac{L^2 \tan \alpha}{2}. \quad \cdot \quad \cdot \quad \cdot \quad \cdot \quad \cdot \quad (2)$$

In general, the actual distance l between the poles is un-known. This term may be eliminated by making another experiment on the deviation of the needle at a different dis-tance.

We get an angle of deviation α', such that

$$\frac{\mathfrak{M}}{\mathfrak{K}}\left(1 + 2\frac{l^2}{L'^2}\right) = \frac{L'^2 \tan \alpha'}{2}; \quad \cdots \quad (3)$$

whence by subtracting (3) from (2), after having multiplied (2) by L^2 and (3) by L'^2,

$$\frac{\mathfrak{M}}{\mathfrak{K}} = \frac{L^2 \tan \alpha - L'^2 \tan \alpha'}{2(L^2 - L'^2)}. \quad \cdots \quad (4)$$

Equations (1) and (4) permit us to determine separately the magnetic moment of the magnet and the horizontal intensity of the earth's field. This apparatus, including the magnetized needle, is called a *magnetometer*, because it enables us to compare the moment of two magnets or the effects of a single magnet placed in different positions.

49. Remarks.—I. The needle might have been hung in a direction normal to the centre of the magnet. Adopting the preceding mode of calculation, we should then have obtained

$$\frac{\mathfrak{M}}{\mathfrak{K}} = \frac{L^2 \tan \alpha - L'^2 \tan \alpha'}{L^2 - L'^2}.$$

II. The results obtained in similar experiments agree rigorously with the calculated results based on Coulomb's law and furnish the proof of the exactness of this law.

50. Measurement of Angles.—The preceding determi-nations necessitate the measurement of the angles of equi-librium of the needle. As this measurement is of frequent occurrence in electrotechnics, we will describe some of the methods of readings in use.

The unit angle employed in absolute measurements is the radian, or arc equal to the radius and corresponding to

$$\frac{360°}{2\pi} = 57°\ 17'\ 44''.$$

The simplest way is to affix to the needle a slender pointer which moves over a horizontal scale. The error of parallax is avoided by placing a mirror in the plane of the scale and putting the eye of the observer in such a position that the pointer coincides with its reflection in the mirror. The reading is thus made in a plane perpendicular to the scale. This mode of reading does not allow of great precision; when the angles are small, the relative error may be considerable.

Greater exactitude is obtained by reflection methods recommended by Poggendorff and Lord Kelvin.

In the first, called *subjective method*, a small plane mirror, *M*, is attached to the axis of suspension of the needle, Fig. 17, and at a certain distance, varying from 3 to 10 feet, is.

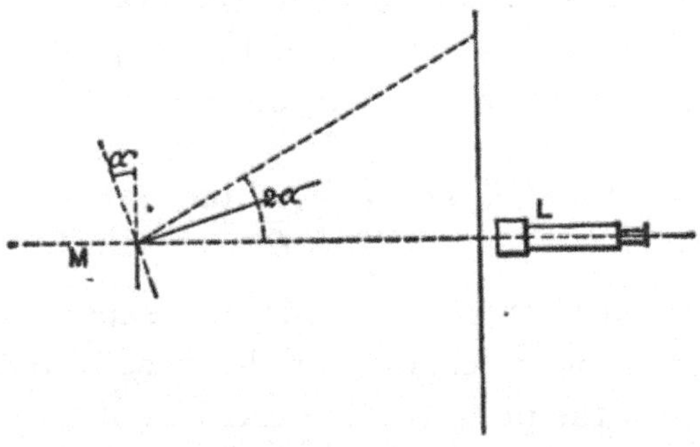

Fig. 17.

placed a reading-telescope with micrometer threads, supported on a tripod with adjustment screws. Above and below the telescope is a horizontal scale, graduated decimally. Before each measurement the apparatus is adjusted to thas.

the vertical plane through the optical axis of the telescope is normal to both the scale and the mirror. To do this, sight at the centre of the mirror and move the telescope and scale until the division of the scale, which lies in the vertical plane passing through the optical axis, is read in the mirror.

The result of this arrangement is, that for an angle of deviation α of the needle, a length l is read off by the tele. scope, such that

$$\tan 2\alpha = \frac{l}{L},$$

L being the distance of the mirror from the scale.

Hence we deduce

$$\alpha = \frac{1}{2}\tan^{-1}\frac{l}{L} = \frac{1}{2}\left(\frac{l}{L} - \frac{1}{3}\frac{l^3}{L^3} + \frac{1}{5}\frac{l^5}{L^5}\cdots\right).$$

If the angles are less than 3°, we need only retain

$$\alpha = \frac{1}{2}\frac{l}{L} - \frac{1}{6}\frac{l^3}{L^3}.$$

This calculation may be avoided by using a scale curved into the arc of a circle of radius L. We then have rigorously

$$\alpha = \frac{1}{2}\frac{l}{L}.$$

In Sir Wm. Thomson's *objective method* the plane mirror is replaced by a curved one, and the telescope by a lamp which emits a ray of light through a diaphragm furnished with a vertical slit. The scale is placed at such a distance from the mirror that the image of the slit is cast on the scale-divisions. These may be marked on a scale made of ground glass, tracing cloth, or celluloid; the observer stands behind the scale and sees the image by transparence. When an incandescent lamp is used, the filament of the lamp gives an extremely precise linear image.

The movable mirror may be plane, as in the preceding case, but a convergent lens must then be placed in the path of the ray of incident light.

INDUCED MAGNETIZATION.

51. Magnetic and Diamagnetic Bodies. — We have seen, § 39, that molecules of iron placed in a field tend to place themselves along the magnetic lines of force. The different varieties of iron, cast iron and steel (with the exception of manganese-steel), together with cobalt and nickel, show an energetic magnetization in a magnetic field. Some other bodies, such as magnetic oxide, perchloride and sulphate of iron, exhibit the same properties, but to a much smaller degree.

Bismuth is also magnetized in a very intense field, but a bar of this metal tends to place itself perpendicularly to the lines of force. In sufficiently powerful fields all bodies exhibit magnetic properties, but to an incomparably smaller extent than iron.

Those bodies whose magnetic orientation is the same as that of iron are called *ferromagnetic* or *magnetic;* those which act like bismuth are called *diamagnetic.*

52. Coefficient of Magnetization or Magnetic Susceptibility. — The problem of magnetization by induction amounts, in fact, to determining for the various parts of the magnets the ratio of the intensity of magnetization to the field-intensity or magnetizing force.

This ratio

$$\kappa = \frac{\mathfrak{J}}{\mathfrak{H}}$$

is called the *coefficient of magnetization* or the *magnetic susceptibility.*

It is easy to see from the dimensions of $\mathfrak{I}$ and $\mathfrak{H}$, §§ 36, 40, that this coefficient simply represents a numerical factor.

If an isotropic body, whose magnetic susceptibility is the same in all directions, is submitted to a magnetizing force that is constant in every point of the body, it tends to acquire a constant intensity of magnetization. . But the magnetic poles induced in the body modify the field, so that if the field were uniform before the introduction of the body, it becomes non-uniform in consequence of that introduction. It is very difficult in the majority of cases to determine the resultant field and, consequently, the actual intensity of the magnetizing force in every point. Thus the problem of the distribution of magnetism in a short cylinder, whose axis is parallel to the direction of the field, has never been solved.

Certain cases, however, are easily calculated. The practical way of obtaining a uniform field of a given intensity consists, as will be seen further on, in sending an electric current through a very long cylindrical bobbin, in the interior of which is placed the body to be magnetized.

53. Cases of a Sphere and a Disk.—Let an isotropic sphere be placed in a uniform field of intensity $\mathfrak{H}$. The various elements of the sphere tend to assume a uniform magnetic orientation, in consequence of which there are developed, on the two hemispheres limited by the great circle normal to the direction of the field, such magnetic shells that the resultant internal action on unit pole is constant and equal to

$$\frac{4}{3}\pi\mathfrak{I}. \quad \ldots \quad \ldots \quad \ldots \quad (\S\ 42)$$

The intensity of the field in the interior of the sphere is therefore constant in magnitude and direction and equal to

$$\mathfrak{H} - \frac{4}{3}\pi\mathfrak{I};$$

consequently

$$\kappa = \dfrac{\mathfrak{I}}{\mathcal{H} - \dfrac{4}{3}\pi\mathfrak{I}},$$

whence

$$.\mathfrak{I} = \dfrac{\kappa\,\mathcal{H}}{1 + \dfrac{4}{3}\pi\kappa}.$$

The susceptibility of iron is always much higher than unity. It follows, then, that the value of $\mathfrak{I}$ is never very distant from $\dfrac{\mathcal{H}}{\dfrac{4}{3}\pi} = \dfrac{\mathcal{H}}{4.19}$. Consequently the spherical form is not suitable for obtaining high intensities of magnetization.

In the interior of an infinitely thin disk, transversely magnetized in such a way that the magnetic density of its faces, equal to the intensity of magnetization, be $\mathfrak{I}$, the component of the force due to these faces is (§ 31)

$$+ 2\pi\,\mathfrak{I} - (- 2\pi\,\mathfrak{I}) = 4\pi\,\mathfrak{I}.$$

Consequently, the intensity of magnetization becomes

$$\mathfrak{I} = \kappa(\mathcal{H} - 4\pi\,\mathfrak{I}),$$

whence

$$\mathfrak{I} = \dfrac{\kappa\,\mathcal{H}}{1 + 4\pi\kappa},$$

a value which tends towards $\dfrac{\mathcal{H}}{4\pi}$.

The transverse magnetization of a disk of iron is therefore always very feeble.

The same can be shown with regard to the magnetization of an iron cylinder in a direction normal to its axis.

54. Case of a Ring.—A ring subjected to magnetizing forces, constant in magnitude and directed in every point of the ring along the tangent to the parallel circle passing through that point, will assume a constant magnetization without free poles, since the rows of magnetic molecules will form closed circular chains.

The original field will not, therefore, have its distribution modified by the presence of the ring, and the intensity of magnetization will be simply expressed by

$$\mathfrak{I} = \kappa \mathfrak{H},$$

$\mathfrak{H}$ representing the intensity of the field.

A conductor coiled round an iron ring, and traversed by an electric current, approximately realizes the above condition, as will be shown later. After the stoppage of the current, the annular core retains the greater part of its magnetism in the permanent state, for, in the absence of free poles, there is no demagnetizing force.

55. Case of a Cylinder of Indefinite Extent.—A third solution is furnished by a cylinder of indefinite extent, placed parallel to the lines of force of a uniform field. The intensity of the field in the interior of the cylinder is the resultant of the original field and of the action of the poles induced at the extremities of the cylinder.

The longitudinal magnetization that can be given to an iron cylinder, whose axis is placed parallel to the direction of the field, increases as the length of the cylinder is increased, since the effect of its poles then produces less and less diminution of the intensity of the field inside the cylinder. Short steel cylinders cannot, therefore, make good permanent magnets, for they become only slightly magnetized, and the reaction of the poles tends to rapidly change the molecular orientation after taking away the magnetizing force.　On the other hand, long cylinders become strongly magnetized and retain their magnetism.

When an iron cylinder is of indefinite length, the demagnetizing action of its poles becomes negligible for points situated in the accessible region of the cylinder where the intensity of magnetization is uniform and expressed by

$$\mathfrak{J} = \kappa\,\mathfrak{K}.$$

It has been shown experimentally that this formula is still applicable when the length of the cylinder is equal to 400 or 500 times its diameter.

56. Portative Power of a Magnet.—Let us consider a cylinder of indefinite length, magnetized parallel to its axis. If we imagine a narrow crevasse cut out normal to the axis, the opposite walls will be covered with magnetic masses whose density is equal to the intensity of magnetization,

$$\sigma = \mathfrak{J}.$$

The force with which unit mass, situated near the face whose density is $-\sigma$, is attracted by this latter, is expressed by $2\pi\sigma$, § 31.

Consequently the mass $+\sigma$, which covers unit surface on the opposite wall, is attracted with a force

$$2\pi\sigma^2 = 2\pi\,\mathfrak{J}^2.$$

This is the portative force of the magnet per unit surface.

If, moreover, the cylinder is under the action of a field of intensity $\mathfrak{K}$,—that is, a field capable of exercising a force $\mathfrak{K}$ on unit pole, and directed parallel to $\mathfrak{J}$,—the portative force must be increased by

$$\mathfrak{K}\,\sigma = \mathfrak{K}\,\mathfrak{J}.$$

The total portative force will then be

$$\mathfrak{K}\,\mathfrak{J} + 2\pi\,\mathfrak{J}^2$$

per unit surface, or

$$(\mathcal{H}\,\mathfrak{I} + 2\pi\,\mathfrak{I}^2)S$$

for a surface S.

These expressions furnish a simple means of determining the intensity of magnetization. In the first case, it is only necessary to apply weights to one of the portions of the cylinder until it is no longer able to sustain them; let w dynes be the weight at which it ceases to sustain,

$$2\pi\,\mathfrak{I}^2 S = w$$

whence

$$\mathfrak{I} = \sqrt{\frac{w}{2\pi S}}.$$

Some authors give the following expression for the portative power :

$$2\pi\,\mathfrak{I}^2 + \mathcal{H}\,\mathfrak{I} + \frac{\mathcal{H}^2}{8\pi}.$$

The last term relates to the force needed to separate the magnetic field itself into two parts. It is the force necessary to separate into two portions an indefinitely long magnetizing bobbin, for we shall see that the mass of the poles of such a bobbin is represented, per unit surface, by $n_1 I$, n_1 being the number of turns per centimetre, and I the current. The portative power of the bobbin is consequently $2\pi n_1^2 I^2 = \frac{\mathcal{H}}{8\pi}$, for $\mathcal{H} = 4\pi n_1 I$. The term $\frac{\mathcal{H}^2}{8\pi}$ is, moreover, very small in the case of an iron core.

57. Variations of Intensity of Magnetization with the Magnetizing Force. Hysteresis.*—When an indefinitely long iron bar, in the neutral state and softened by annealing, is subjected to the influence of a field of increas-

---

* See also chapter on Hysteresis, by Charles Proteus Steinmetz, p. 87.

ing intensity, the magnetization $\mathfrak{I}$ varies as shown by the curve *OA*, Fig. 18, whose abscissæ represent the values of $\mathfrak{K}$. We see that for very small forces the magnetization increases slowly. Beyond $\mathfrak{K} = 1$ C. G. S. unit, the curve shows a point of flexion beyond which the ordinates grow rapidly larger up to a point corresponding to values of $\mathfrak{K}$ comprised between 5 and 10 C. G. S. units, where the curve makes a sharp turn. The increase of the ordinates then becomes smaller and smaller and the bar reaches the state commonly known by the name of *saturation*, which corresponds rigorously to the ordinate of a horizontal asymptote to which the curve approaches indefinitely. The magnetizing forces

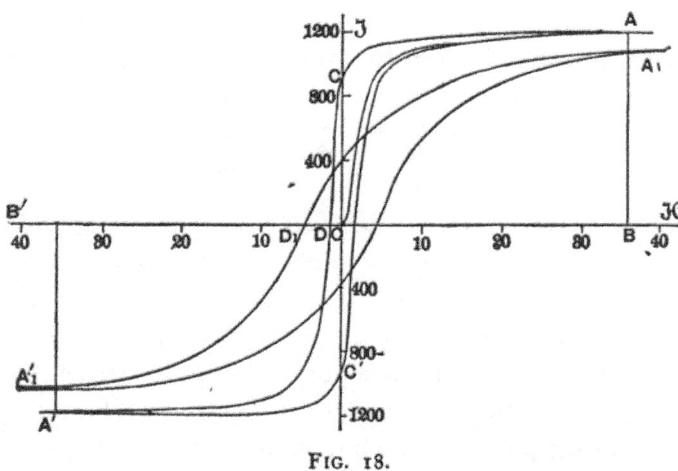

FIG. 18.

shown here are obtained in practice by placing the bar in a very long solenoid traversed by an increasing current. The intensity of the field inside the solenoid is proportional to the current.

According to Ewing and Low, the intensities of magnetization corresponding to saturation are approximately, in C. G. S. units, for wrought iron 1700, for cast iron 1240, and for nickel 513.

Steel, which can attain the same magnetization as iron under very powerful magnetizing forces, scarcely retains more than half in the form of permanent magnetism.

The magnetization-curve shows that the susceptibility $\kappa = \dfrac{\mathfrak{J}}{\mathfrak{K}}$ is at first very feeble; then it increases rapidly with $\mathfrak{K}$ and attains for iron a value varying from 200 to 300. It then decreases progressively to a very small value.

For slightly magnetic bodies the susceptibility is always less than 0.00001 in absolute value; in these conditions it may practically be considered as zero compared to the susceptibility of iron, for moderate magnetizing forces.

Let us suppose that the bar, after having reached the point A corresponding to saturation, be submitted to magnetizing forces decreasing from OB to zero. The magnetization does not pass again through the intermediate states first observed, but varies according to the curve AC, OC corresponding to the residual magnetism.

If the magnetizing force changes its direction and takes a negative value OB', equal to OB, the intensity takes the successive values shown by the curve CA'. Finally, the magnetizing force repassing through the consecutive values between B' and B, the magnetism of the bar will return to the value AB by a curve $A'C'A$.

The cycle $ACA'C'A$ can be reproduced indefinitely by causing the field-intensity to vary periodically between the values OB and OB', which is obtained by giving to the current traversing the magnetizing solenoid values oscillating between two equal limits of contrary sign. The curves joining the points A_1, A_1', Fig. 18, show the variations of the magnetic state of the same iron bar hardened by the application of a tractive force greater than the limit of elasticity of the metal. It will be observed that the maximum magnetization is less than when the metal is annealed.

Moreover, the susceptibility of the hardened metal is con. siderably diminished.

It will be seen by the above that the magnetization of a bar is capable of assuming very different values for the same magnetizing force; it depends not only on the actual mag. netizing force, but also on the preceding magnetic condi. tions. The intensity of magnetization of an iron core is a complex function of the magnetizing force and the preceding condition of the iron.

During the period of decrease in the cyclic curve, the values of the intensity of magnetization are always larger than those given by the curve OA, while during the period of increase they are smaller. This phenomenon, due to the coercive force, has been called by Ewing *hysteresis* (from the Greek, *lagging behind*).*

The ordinate at the origin, OC, represents the residual magnetism of the bar. When soft iron is kept from all vibrations, this ordinate equals nearly three fourths of the maximum ordinate of the curve.

Dr. Hopkinson has especially designated by the name *coercive force* the magnetizing force, OD, which must be applied to the bar (in reverse direction) to destroy its residual magnetism. The bar is then not, however, in the neutral state, for it has a very different susceptibility from that observed on beginning the magnetization; for it is readier to take a negative magnetization than when it was in the neutral state. In the hardened metal the coercive force, $OD_{,}$, is appreciably increased.

The form of the curve $ACA'C'$ shows that to bring a bar back to the neutral state it must be subjected to periodic forces of decreasing intensity. The cycles then described will approach nearer and nearer to the origin. It is for this reason

---

* See Ewing, *Magnetic Induction in Iron, etc.*, p. 93 *et seq.*

that to demagnetize a watch it must be placed in the field
of a magnet and then steadily drawn away, making it at
the same time revolve in order to change the direction of
the magnetizing force, which grows feebler as the distance
from the magnet increases.

Figure 19 is intended to show the difference existing between the neutral state and that of zero magnetization. A
bar in the neutral state has been subjected to increasing
magnetizing forces, bringing it up to the magnetization *A*.
The field has then been gradually reduced to zero and next
changed in direction. At a certain point, a momentary

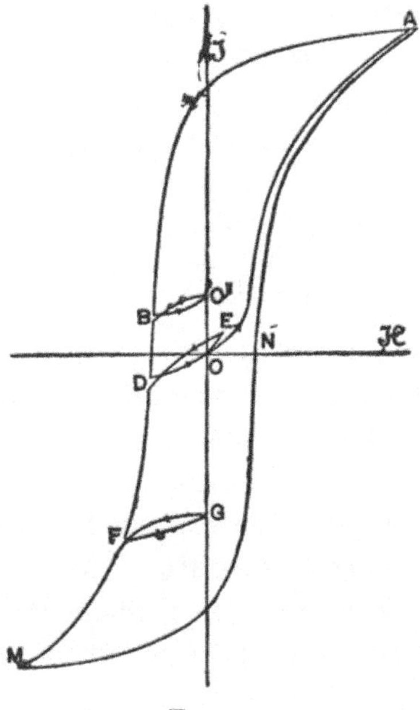

FIG. 19.

return of the magnetizing force to zero has produced the
loop *BC*; then the values of the field have recommenced to
decrease. A new retrogression has occurred at *D*, chosen
so that the curve of magnetization passes through the

origin.   At the moment of passing through the origin, the bar is not in the neutral state, for if the magnetizing force be increased, the curve continues along OE and not along OA.   On completing the cycle of the field-intensity, we return to a point which does not coincide with A unless this latter corresponds to the point of saturation of the bar.

58. Frölich's Formula.—Various writers have tried to represent, by empirical formulæ, the variation of the intensity of magnetization as a function of the magnetizing force.

If we suppose that the susceptibility is proportional to the difference between the maximum intensity of magnetization and the actual intensity, we have

$$\kappa = \frac{\mathfrak{I}}{\mathfrak{K}} = A(\mathfrak{I}_m - \mathfrak{I}),$$

whence

$$\mathfrak{I} = \frac{A\,\mathfrak{I}_m\mathfrak{K}}{1 + A\mathfrak{K}} = \frac{a\,\mathfrak{K}}{1 + b\,\mathfrak{K}},$$

a and b being constants for a given bar.

This curve represents an hyperbola passing through the origin and one of whose asymptotes is parallel to the axis of x.

By choosing the parameters a and b suitably, we can, for approximate calculations, substitute this curve for the true one obtained by experiment.   Frölich and S. P. Thompson have applied this formula to the theory of dynamos.

59. Müller, von Waltenhofen, and Kapp have adopted a formula of the form

$$\mathfrak{I} = a \tan^{-1} b\,\mathfrak{K},$$

which can likewise furnish approximate values of intensities of magnetization of soft iron by a suitable choice of parameters. This formula is not so convenient in calculating as the preceding one. It will be noticed that the above equations represent curves passing through the origin and that they consequently leave out the phenomenon of hysteresis. They give, at the most, a curve intermediate between the two curves obtained in a magnetic cycle.*

60. Another Way of Looking at Induced Magnetization, Magnetic Induction, and Permeability. — Let us consider a uniform field, in which the intensity $\mathcal{H}$ represents the flux of force across unit equipotential surface, which is measured by the force exercised on unit pole. If we place an indefinitely long cylinder parallel to the direction of the field, the space occupied by the cylinder becomes the seat of a different flux, that is, the force exercised on a unit pole, hypothetically placed inside the cylinder, is modified in a way that will appear later on. This flux, $\mathcal{B}$, per unit section is sometimes called the *magnetic induction* across the cylinder.

---

* Drs. Houston and Kennelly have recently pointed out that from the researches of Ewing, Klaassen, Fessenden, and others

(1) An approximate linear relation exists between remanance and maximum cyclic intensity in iron and steel, or, symbolically,

$$\mathcal{B}_{0} = a + b\,\mathcal{B}_{max}.$$

where $\mathcal{B}_{0}$ is the remanance or residual cyclic magnetic flux when the magnetizing force is zero, and $\mathcal{B}_{max}$. is the maximum cylic intensity. This can only be regarded as an empirical formula, holding between $\mathcal{B}_{max}. = 500$ and $\mathcal{B}_{max}. = 9000$, in some cases up to $\mathcal{B}_{max}. = 16{,}000$.

(2) An approximate linear relation exists between coercive force and maximum cylic intensity in iron and steel; or, symbolically,

$$\mathcal{H}_{0} = a_{1} + b_{1}\mathcal{B}_{max}.$$

where $\mathcal{H}_{0}$ is the cyclic coercive force or the value of $\mathcal{H}$ at which $\mathcal{B} = 0$, and $\mathcal{B}_{max}$. is the maximum cyclic intensity. This can only be regarded as an empirical formula, holding above $\mathcal{B}_{max}. = 4000$.

(3) As a consequence of the preceding relations an approximate linear relation exists between remanance and coercive force in the cyclic magnetization of iron and steel between the limits above defined.

The ratio

$$\mu = \frac{\mathfrak{B}}{\mathfrak{H}} \quad . \quad . \quad . \quad . \quad . \quad . \quad . \quad . \quad . \quad (1)$$

between the magnetic induction and the magnetizing force is the *coefficient of permeability* (or simply *permeability*) of the cylinder. It follows from this definition that the permeability of the medium into which the substance is introduced, generally the air, is taken as unity.

The permeability depends on the nature of the substance and, in highly magnetic bodies, on the intensity of the field. The values of $\mathfrak{B}$ and μ may be determined directly by electric measurements. These quantities are rendered important by the fact that they are connected by a simple relation with the intensity of magnetization and the susceptibility.

In order to estimate the flux of force inside the cylinder, let us suppose, as in § 56, that an infinitely narrow crevasse be cut in the cylinder perpendicularly to its axis. We can consider that this operation does not modify the total flux across the cylinder. The walls of the crevasse normal to the direction of the cylinder are charged by induction with free magnetism having densities $+ \sigma$ and $- \sigma$, such that

$$\sigma = \mathfrak{I},$$

$\mathfrak{I}$ being the intensity of magnetization of the cylinder.

The effect of these surface-charges on unit pole, supposed to be introduced into the middle of the crevasse, is to produce two components in the same direction, equal to

$$2\pi\sigma = 2\pi\,\mathfrak{I},$$

having a direction parallel to the axis of the cylinder, § 31. Besides this we must add to these components the force $\mathfrak{H}$ due to the field. Since the two components are parallel by hypothesis, the resultant will be

$$\mathfrak{B} = \mathfrak{H} + 4\pi\mathfrak{I} = \mathfrak{H}(1 + 4\pi\kappa). \quad . \quad . \quad . \quad . \quad (2)$$

Comparing equations (1) and (2), we see that

$$\mu = 1 + 4\pi\kappa. \quad . \quad . \quad . \quad . \quad . \quad (3)$$

It follows from the preceding equations that we can express the magnetization of a body in a field by the intensity of magnetization or by the magnetic induction indifferently. It seems at first sight that one of these expressions is superfluous, and that the use of both can only produce a confusion of ideas. But, as will be seen more clearly further on, there are cases where it is more convenient to use the first expression, and other cases where the second expresses the phenomena more clearly. Thus, when we consider a magnetized bar, the moment of which is determined by the magnetometer, § 48, the intensity of magnetization is expressed by the ratio of its moment to its volume. But in the case of a ring, § 54, in which the lines of force are closed, the external magnetic effect would be zero, as would the moment also, for the ring has no free pole and its magnetism cannot be called into action except by making a section in a plane passing through the axis of revolution. The walls thus exposed present poles of contrary name and whose density represents the intensity of magnetization of the body. Between these poles a uniform field is developed in which there is a flux equal to $4\pi\mathfrak{I}$ per unit of section and which is to be added to the flux $\mathfrak{H}$ in the same direction due to external causes. It will be seen in the part on Electromagnetism that the total flux, called magnetic induction across the ring, is capable of being determined directly. The intensity of magnetization is deduced from this quantity by subtracting the value of $\mathfrak{H}$ and dividing the remainder by 4π.

Some authors estimate the magnetism of magnetized bars

in units of magnetic induction. In this case it is only necessary to multiply by 4π the mean value of the intensity of magnetization found by means of the magnetometer.

While, in the case of a ring, the flux of magnetic force remains in the iron, in the case of a straight magnet the flux leaves the iron and is closed through the surrounding air. In the first example the medium is homogeneous, in the second heterogeneous, being composed partly of iron and partly of air; but in both cases the flux should be considered as continuous and closed on itself. This way of looking at the matter has helped to simplify the conception of magnetic phenomena. It will appear in the course of this work how much has been gained by extending to circuits traversed by magnetic fluxes the conditions shown to exist for circuits traversed by electric currents.

By definition, the permeability of air is unity. Experiments show that the permeability of a vacuum is sensibly of the same value. Magnetic bodies are those of which the permeability exceeds that of air; diamagnetic bodies, those of which the permeability is less than that of air. This can also be expressed in the statement that magnetic bodies conduct lines of force more readily, and diamagnetic bodies less readily, than air.

From the relation

$$\mu = 4\pi\kappa + 1$$

we have

$$\kappa = \frac{\mu - 1}{4\pi}.$$

This shows that the susceptibility of magnetic bodies for which $\mu > 1$, is superior to zero, while the susceptibility of diamagnetic bodies is negative.

The susceptibility and permeability of iron, cobalt and nickel at ordinary temperatures are so superior to those of other bodies, whether magnetic or diamagnetic, that there is no practical error in taking the permeability of all other bodies as equal to unity and their susceptibility as zero. The most diamagnetic body in existence, bismuth, has a permeability of 0.9991.

61. Work spent in Magnetizing.—As the magnetization of a body imparts to it a certain quantity of potential energy, it necessitates a certain expenditure of work. We shall show later on that this work is expressed, per unit volume of the magnetized body, by

$$\frac{1}{4\pi}\int \mathfrak{H}\, d\mathfrak{B} = \frac{1}{4\pi}\int \mu\,\mathfrak{H}\, d\mathfrak{H},$$

the integral being extended to the limits between which the induction of the magnet has been changed. If the values of the induction compared with the field-intensity be shown by the curve OA, Fig. 20, and if the magnetization reach the state denoted by the point A, the work expended will

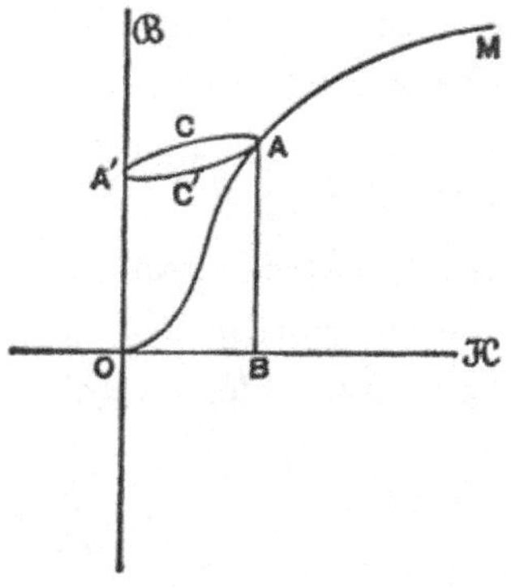

FIG. 20.

be represented by the area, divided by 4π, of the surface comprised between the curve, the axis of ordinates and a parallel to the axis of abscissæ drawn through A.

If μ were a constant factor, as is the case for slightly magnetic substances, the integral would reduce to

$$\mu\,\frac{\mathfrak{K}^2}{8\pi},$$

for a variation extending between o and $\mathfrak{K}$.

Let us suppose that an indefinitely long bar, after having reached the magnetic state A, traverses a cycle $ACA'C'A$, Fig. 20, the intensity of the field passing from the value OB to o, and then returning to OB.

The integral

$$\frac{1}{4\pi}\int_{\mathfrak{B}\,=\,OA'}^{\mathfrak{B}\,=\,AB}\mathfrak{K}\,d\mathfrak{B}\ ,$$

representing the area of the surface $A'C'AC$ divided by 4π, expresses the work expended in order to cause unit volume of the bar to pass through the given cycle. This work is transformed into heat in the substance, and is the loss due to hysteresis.

In the case where the cycle through which the magnetized bar goes is produced by forces oscillating between values OB and OB', Fig. 21, the loss by hysteresis is represented by the area $ACA'C'A$.

In the case of a completed cycle the expression for the energy dissipated is susceptible of a simpler expression. Thus, replacing $\mathfrak{B}$ by $4\pi\,\mathfrak{I} + \mathfrak{K}$, we get

$$\frac{1}{4\pi}\int\mathfrak{K}\,d\mathfrak{B} = \int\mathfrak{K}\,d\mathfrak{I} + \frac{1}{4\pi}\int\mathfrak{K}\,d\mathfrak{K}.$$

Now the second integral is cancelled for a completed cycle and the energy is then expressed by

$$\int\mathfrak{K}\,d\mathfrak{I}.$$

It is readily seen that the loss increases with the coercive force, represented by the abscissa at the origin of the curve *ACA'*, and with the maximum intensity of magnetization.

When a magnet is in a state of repose the loss is greater than in the dynamic state, especially for soft iron, for we have seen that the coercive force is diminished by vibrations of the substance.

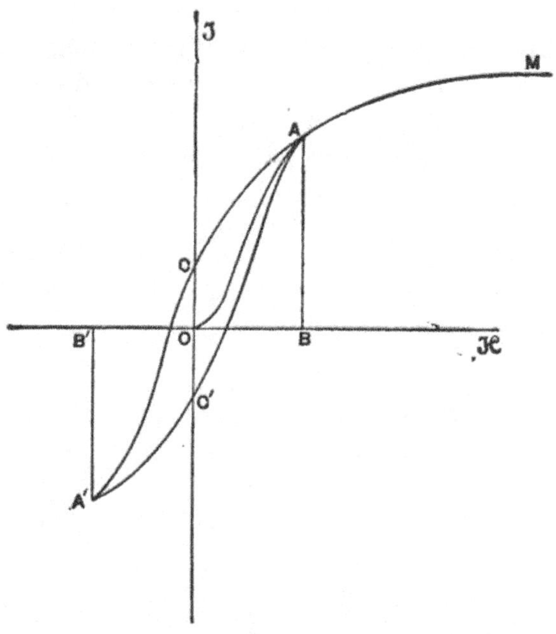

FIG. 21.

When the cycle of the magnetizing force is completed very rapidly, the magnetization attained by the metal is diminished, as well as the loss by hysteresis. Thus M. Tanakadate has found that when the duration of the cycle is between $\frac{1}{28}$ and $\frac{1}{400}$ of a second, the loss is only eight tenths of that observed in the case of cycles completed more slowly.

62. Numerical Results.—The preceding equations show that the values of the permeability of an iron bar pass through variations analogous to those of its susceptibility. At first very small for small values of the magnetizing force, the permeability grows rapidly towards a maximum, then decreases indefinitely towards a value but slightly differing from that of air.

The curves, Figs. 18 and 19, showing the variations of intensity of magnetization of a bar in terms of the magnetizing force, also show, very sensibly, the magnetic induction referred to this same force if we regard unity on the scale of the ordinates as representing the number 4π.

Below is a table of magnetic values as found for two specimens of annealed soft iron, and one of gray cast-iron.

Annealed Soft Iron.				Gray Cast-iron.	
$\mathfrak{B}$	μ	$\mathfrak{B}$	μ	$\mathfrak{B}$	μ
1,000	560	11,000	1,692	4,000	800
2,000	880	12,000	1,412	5,000	500
3,000	1,160	13,000	1,083	6,000	279
4,000	1,400	14,000	823	7,000	133
5,000	1,600	15,000	526	8,000	100
6,000	1.800	16,000	320	9,000	71
7,000	1,960	17,000	161	10,000	53
8,000	2,120	18,000	90	11,000	37
9,000	2,280	19,000	54		
10,000	2,000	20,000	30		

Very pure soft iron shows the greatest permeability of any metal. It is followed, in descending order, by soft steels (Thomas and Bessemer), malleable iron, and gray cast-iron, the magnetic qualities of which are very variable according to its composition. Tempered steel, in which the iron is in an especial condition, has a very low permeability, while steel containing 12 per cent. of manganese is hardly more magnetic than air.

The following curves represent the permeability as a function of the magnetic induction for various specimens of iron experimented on by Rowland, Hopkinson, and Bidwell.

The coercive force, measured by the abscissa at the origin of the curve of magnetism, is only about 2 for soft iron; it

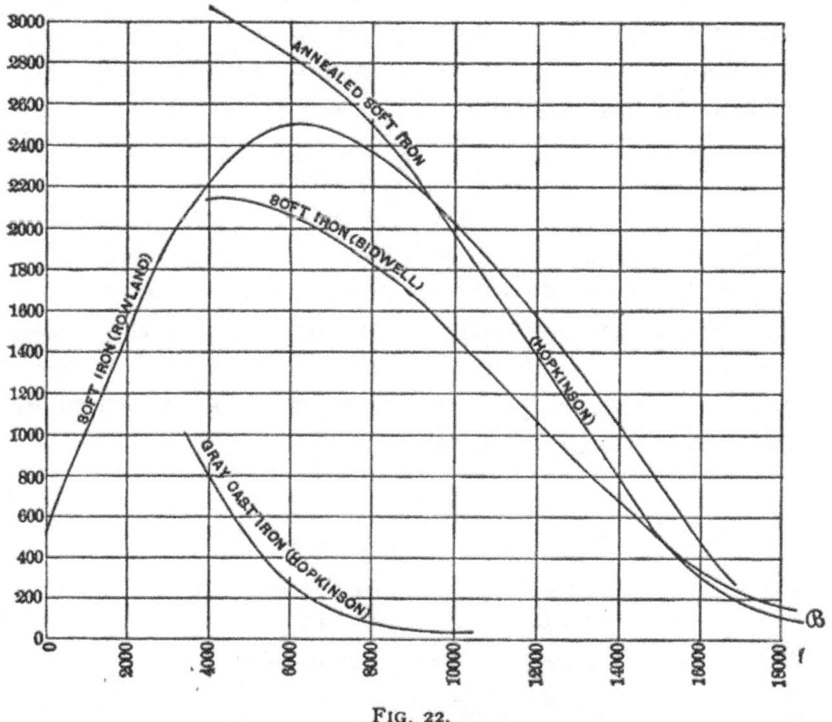

FIG. 22.

reaches 40 for chrome-steel tempered in oil, and 50 for steel containing 3 to 4 per cent. of tungsten. This result shows that the latter steel is especially suitable for making permanent magnets.

The energy dissipated by hysteresis is in proportion to the maximum induction to which the metal is subjected and to its coercive force. While this energy varies between

10,000 and 15,000 ergs for specimens of iron and soft steel subjected to magnetizing forces oscillating between values high enough to produce saturation, it attains 216,000 ergs in tungsten-steel (Hopkinson).

The permeability of annealed soft iron decreases very nearly in proportion to the increase of $\mathfrak{B}$, between values corresponding to $\mathfrak{B} = 7$ kilogausses and $\mathfrak{B} = 16$ kilogausses. Between these limits the mean values of the permeability are approximately given by the empirical formula, deduced from Hopkinson's curve,

$$\mu = -\frac{\mathfrak{B}}{3 \cdot 5} + 4850.$$

Figure 23 presents a curve determined by Ewing and showing the loss in a soft-iron bar subjected to increasing alternating magnetizing forces. The ordinates of the curve denote ergs per cm³ and the abscissæ give in C. G. S. units the extreme values, positive and negative, of the magnetic induction through the metal.

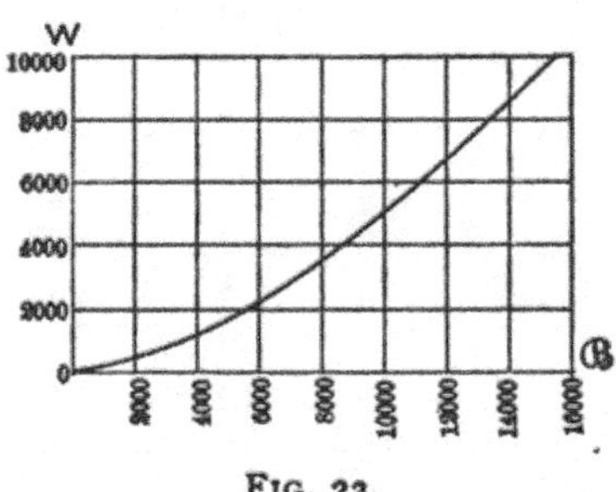

FIG. 23.

According to Steinmetz, the loss of energy in ergs is represented by the expression*

$$w = \eta \, \mathfrak{B}^{1 \cdot 6},$$

---

* Steinmetz, *L'Industrie électrique*, March 6, 1892.

where η, the *coefficient of hysteresis*, may have the following values:

Material.	Composition and State.	Coefficient of Hysteresis. η
Soft iron...............	Annealed	.00202
Soft Bessemer steel.....	.045 per cent. of carbon, annealed	.00262
Whitworth steel........	.032 " " " "	.00598
" "	.032 " " " " tempered	.00954
Manganese-steel........	4.73 " " " manganese, forged	.05963
Tungsten-steel.........	3.35 " " " tungsten, tempered	.05778
Gray cast-iron..........	3.45 " " " carbon	.01826

With the magnetizing forces that can be obtained in dynamo-electric machines, their iron cores seldom exceed an induction of 20,000 C. G. S. units, or gausses, but by establishing particularly powerful fields Messrs. Ewing and Low have succeeded in communicating to very soft iron an induction of 45 kilogausses. Under high inductions, the intensity of magnetization has a constant value of about 1700 C. G. S. units, corresponding to saturation, and the permeability falls to a constant value of between 1 and 2.

We then get the relation

$$\mathcal{B} = \mathcal{H} + 4\pi\,\mathcal{I} = \mathcal{H} + \text{constant.}$$

In very intense fields cobalt is capable of attaining the same maximum intensity of magnetization as cast-iron, or about three fourths of the magnetization of soft iron. Nickel never exceeds one third of the maximum intensity of magnetization of soft iron.

Lord Rayleigh has found that in very weak fields the permeability can be expressed by a formula

$$\mu = a + b\,\mathcal{H}.$$

For a specimen of soft iron he has found $a = 81$ and $b = 64$.

He has also established the fact that hysteresis is absent

when a bar is subjected to magnetizing forces varying between very narrow limits, whether the metal already possesses
any magnetization, or if it has been taken in the neutral
state; in these conditions the permeability is constant.
When these small variations occur near a magnetizing force
of 29 C. G. S. units or gilberts, he has found that the permeability of soft iron is only 80 per cent. of the permeability
near the neutral state.

**63. Effect of Temperature on Magnetism. Recales-
cence.**—We have already made allusion to the influence of
the temperature on the magnetism of iron and its derivatives, steel and cast-iron, whose magnetism disappears completely at a bright red heat.

Dr. Hopkinson, to whom we are indebted for precise
experiments on this thermic effect,* has observed that in a
feeble and constant field of 0.3 gauss the permeability of
a soft-iron bar, heated gradually, increases progressively
from 500 to 11,000; but at the temperature of 775° C., the
permeability falls suddenly to a value very close to 1.

When the intensity of the field increases, the increase of
permeability is much less sensible and the fall is less sudden.
Finally, in an intense field, the permeability decreases continuously with the rise of temperature. In every case the
iron becomes completely demagnetized at a temperature
in the neighborhood of 785° C.; Dr. Hopkinson calls this
thermic point the *critical temperature* of the metal.

For exceptionally soft iron the critical temperature may
rise as high as 880° C., while in steel it falls to 690°. For
nickel the critical temperature is about 310° C.

The critical temperature seems to correspond to a molecular change in the substances, shown likewise by other

---

* See Hopkinson, *Magnetism, Journal of the Institution of Electrical
Engineers*, vol. xix.

phenomena. Kohlrausch has observed that at this temperature the electrical resistance of iron shows a sudden variation.

According to Tait the thermo-electric power of iron is also modified in a profound degree towards this point ; and lastly, Barrett has discovered a very characteristic effect, to which he has given the name of *recalescence.* If we allow a piece of iron or steel to cool down after having heated it to a bright red, there comes a certain stage where the process of cooling stops and where the piece becomes slightly heated again, after which the decrease of temperature goes on again regularly. This recalescence is shown in hard steel by a very visible luminous effect, the color of the metal passing from a dull red to very bright red at the moment when the critical temperature is reached. This experiment succeeds very well when a knitting-needle is used, first heating it to a bright red by passing an electric current through it.

It is very surprising that magnetic qualities should be clearly exhibited by only three metals—iron, nickel, and cobalt. The other elementary bodies are so little capable of magnetization that they are ordinarily considered as non-magnetic. It may be that it is only a mere question of temperature, the three metals mentioned being the only ones which manifest decided magnetic properties at the ordinary temperatures. This was Faraday's opinion, who thought that all substances would become magnetic at a sufficiently low temperature. The following fact discovered by Dr. Hopkinson seems to support this opinion: An alloy of iron containing 25 per cent. of nickel is non-magnetic like all alloys. But if this alloy be cooled to slightly below 0° C., it is capable of becoming magnetized in a very marked degree. It possesses, therefore, a low critical temperature. If the alloy be afterwards reheated, it remains magnetic and its susceptibility increases up to about 525° C., at this point the susceptibility falls rapidly and becomes zero at 580° C. Upon recooling the metal, it does not reassume its susceptibility until below 0° C.

64. Ewing's Addition to Weber's Hypothesis.*—In order to explain in Weber's hypothesis of the molecular constitution of magnets, § 39, the coercive force and the loss due to hysteresis, it has been supposed that the elementary magnets (or magnetic molecules) offer a resistance to orientation in the nature of friction and that it is the work spent in overcoming this friction which constitutes the loss by hysteresis. The existence of such a passive resistance enables us, up to a certain point, to account for the effect of vibrations and temperature on magnets, but it does not at all explain the changes in susceptibility especially shown in the regions *A*, *B* and *C* of the magnetism-curve, Fig. 24.

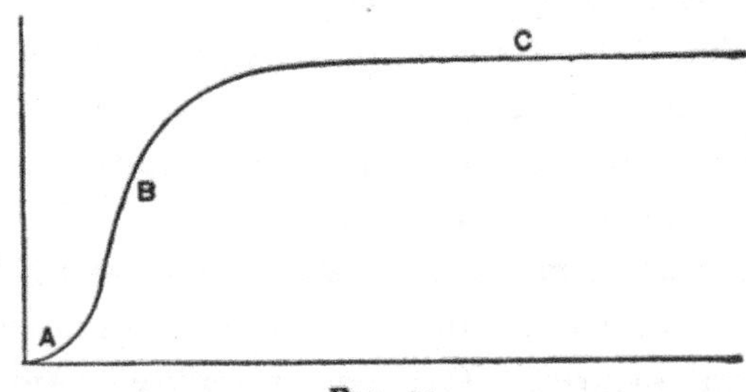

FIG. 24.

Ewing has found experimentally that the observed phenomena are to be explained without bringing in the supposition of friction, by the simple effect of the mutual reactions of the elementary magnets. He has reached this conclusion by investigating the way in which a system of magnetic needles acts, when they are arranged regularly one next the

---

* Ewing *Contributions to the molecular theory of induced magnetism*, Roy. Soc. 1890 ; also *Magnetic Induction, etc.*, pp. 287–8 *et seq.*

other so as to be able to oscillate in the same horizontal plane without touching each other. These needles are subjected to a magnetizing force obtained by rolling coils of wire, carrying an electric current, around the case enclosing them. When the needles are left to their own reactions, that is, when the field produced by the current neutralizes the earth's field, it is observed that they form more or less complex geometrical combinations with each other in stable equilibrium. If one of the elements of this combination is slightly altered from its position, it immediately returns to it; but if the alteration of position is considerable, the combination is not formed again, and new combinations are formed between the neighboring magnets. If a progressively increasing directive force is applied to such a system, it is seen that the various combinations are at first slightly deformed without being destroyed. This, which might be termed an elastic deformation since it is reversible by withdrawing the magnetizing force, is comparable to the state of the molecules of a magnetized bar in the region A of the magnetism-curve, Fig. 24.

If the current through the coils is continuously increased, a point is reached where one of the combinations of the magnets exceeds the limiting deformation which it can stand. There is then produced a sudden change in this combination, and, by the mutual action, the whole system enters on a state of unstable equilibrium, so that a very slight increase in the directive force is sufficient to align all the magnets in a direction approaching to that of the directive force itself. This period corresponds to the region B of the magnetism-curve. The observation of the propagation of the new groupings from one group to the next one is eminently suggestive as explaining the necessity of a definite interval of time for the molecules of a magnet to assume their positions of equilibrium under the action of a magnet-izing force.

When, after this, we still continue to apply increasing forces, we observe that the mutual reactions of the magnets are more and more overpowered and that they align them-selves in a direction which eventually coincides with that of the acting field. This is the state designated under the name of saturation and shown at *C* in the curve.

Figures 25, 26, and 27 show three successive states of the system of magnets; Fig. 25 corresponding to the end of state *A*, Fig. 26 to the end of state *B*, and Fig. 27 to the end of state *C*. If we diminish the intensity of the field, the magnets still maintain their general orientation, but become slightly displaced by the effect of their own reactions. When the directing field becomes zero, the magnets remain more or less aligned in the direction of the field, which accounts for the residual magnetism. But if the directing field changes sign and increases in the opposite direction, we soon observe a sudden return to the state of unstable equilibrium, and then an orientation in the opposite direc-tion.

According to the curve *ACA'*, Fig. 18, annealed soft iron is in a state of unstable equilibrium when the magnetizing force becomes zero, since the elbow of the curve is on the side of the positive magnetizing forces. With tempered steel, on the contrary, this elbow is produced on the side of the negative abscissæ, which accounts for the aptitude of this metal for retaining its residual magnetism. In fact, at the moment when the magnetizing force becomes zero, the state of the metal is shown by a point of the curve situated in the region corresponding to stable equilibrium.

If, instead of placing the magnets regularly, we arrange them at varying distances, we observe that the duration of the state corresponding to unstable equilibrium (*B*) is increased. This is observed in cold-hammered iron, the arrangement of whose molecules has been changed by mechanical means. The various combinations of elements which are formed in such a case are more independent of each other than in the homogeneous metal, and one or another of them can be modified without influencing the adjoining groupings.

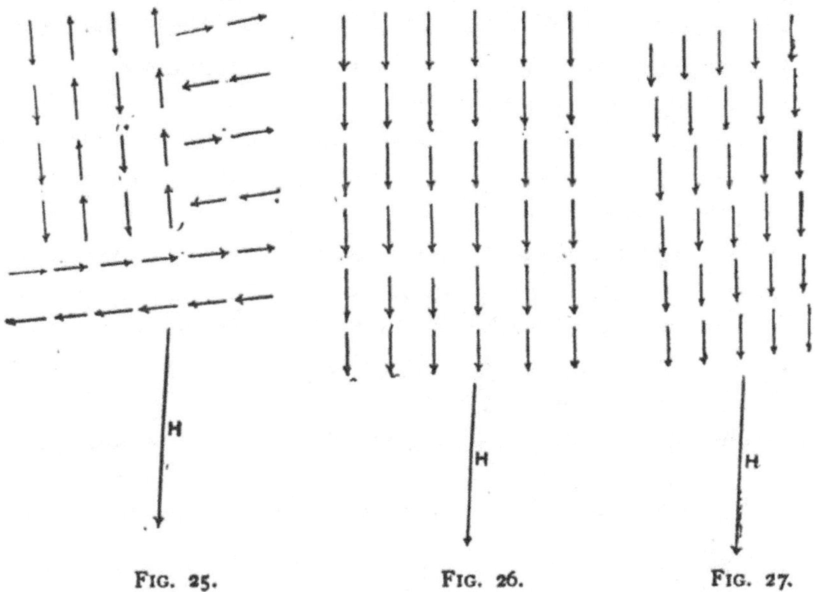

FIG. 25.   FIG. 26.   FIG. 27.

In steel and cast-iron, where the stage B is of great extent, the groupings of the molecules are influenced by the presence of foreign bodies.

According to Ewing, the heating due to hysteresis corresponds to the oscillations of the magnets on passing from one position of stable equilibrium to another.

Vibrations diminish the stability of the combinations and consequently facilitate the orientation of the magnets placed under the action of the field, as well as their return to the neutral state when the magnetizing force has ceased to act.

A rise in temperature produces analogous effects when the magnetizing force is feeble; we have already seen, however, that heating reduces the permeability when the magnetizing force is intense. Ewing explains this fact by considering that the molecular agitation caused by a rise in temperature corresponds to oscillations of the magnets about their axes. When the magnets are already oriented, these oscillations result in a diminution of the mean external action of the system. Or we can admit, with Dr. Hopkinson, that the magnetic moment of the elementary magnets decreases when the temperature increases.

Lastly, the absence of hysteresis, observed by Lord Rayleigh in the case of very feeble variations in the magnetizing force, can be explained if we observe that such variations produce only feeble displacements of the elementary magnets about their positions of equilibrium; such displacements are reversible without break of equilibrium and consequently without the extensive movements which give rise to the development of heat.

The irreversible variations, which are made evident by the separation of the ascending and descending curves of magnetism, are the only ones which give rise to the evolution of heat.

65. Equilibrium of a Body in a Magnetic Field.— We have seen, § 45, that a shell free to move in a magnetic field moves so that the flux entering by its negative face may be a maximum. This conclusion extends to any magnetized body whatever which may be considered as formed by superposed shells.

Thus in a uniform field the axis of an iron cylinder of elongated form aligns itself parallel to the lines of force of the field, in such a way that the flux of force enters by the induced south pole.

It has been shown, §§ 53, 55, that this position corresponds to a more intense magnetization of the metal than does any other position. In a uniform field an isotropic sphere is in equilibrium in all positions, while an anisotropic sphere, whose permeability varies in different directions, orients itself so that the direction of the field may be parallel to the axis of maximum permeability.

The same considerations show that in the neighborhood of a magnet where the field is variable, magnetic bodies tend to move towards the poles so that the flux traversing them may be a maximum.

These movements are clearly exhibited in a liquid placed in a watch-glass over the poles of a powerful electro-magnet.

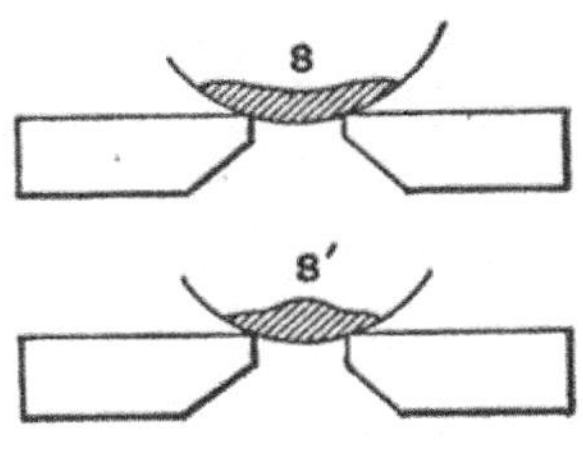

Fig. 28

A solution of sulphate of iron, S, Fig. 28, will present a concavity towards the centre.

Diamagnetic bodies, on the other hand, appear to be repelled by the poles; thus a bar of bismuth assumes a position at right angles to an electro-magnet; a solution of bisulphide of carbon, S', Fig. 28, is heaped up in the middle of the vessel.

Becquerel and Faraday have found a simple explanation of diamagnetic repulsion, by means of comparing it with the action of gravity upon bodies plunged in a liquid denser than themselves. These bodies are apparently repelled by the earth; in the same way it may be that the action of magnets on diamagnetic bodies is simply due to the fact that they are less magnetic than the air or the medium which surrounds them. From this point of view there would be, properly speaking, no diamagnetic substances, but only degrees of permeability. This hypothesis, combatted by Tyndall, has been recently confirmed by Messrs. Parker and Duhem.

HYSTERESIS AND MOLECULAR MAGNETIC FRICTION.*

66. Hysteresis.—Some materials, such as iron, nickel, etc., when exposed to the action of a magnetomotive force, that is, when in a magnetic field, have induced in them a magnetic flux far in excess of that set up under the same conditions in air or other materials. The former are therefore called magnetic materials.

In air and other non-magnetic materials the magnetic flux, $\mathfrak{B}$, varies proportionally to the magnetomotive force, $\mathfrak{F}$, or to the field intensity, $\mathfrak{H}$. In magnetic materials, such as iron, the magnetic flux is proportional to the M. M. F. only for very low values of the latter. With increasing M. M. F.'s it begins to increase at a greater rate than the M. M. F., becomes proportional again to it at still higher values, and for very high M. M. F.'s increases more and more slowly until ultimately its further increase with increase of M. M. F. approaches a limit where it is not greater than in non-magnetic materials; or, in other words, the difference $\mathfrak{B} - \mathfrak{H}$ approaches a finite limit, called the *absolute magnetic saturation* of the material,† which in soft iron corresponds to about $\mathfrak{B} = 20{,}000$, in nickel to $\mathfrak{B} = 6000$, in cobalt to $\mathfrak{B} = 13{,}000$, etc.

The curve indicating the variation of the magnetic flux $\mathfrak{B}$ with the M. M. F. or field intensity is of a form shown

---

* By Chas. Proteus Steinmetz.

† "Magnetic Reluctance," by Kennelly. *Transactions* of the American Institute of Electrical Engineers, 1891.

in Fig. 29 by the full line. This curve is obtained if the M. M. F. acting upon the iron is made to gradually increase from zero to a maximum. If now the M. M. F. is gradually reduced again from the maximum to zero, the corresponding values of magnetic flux, ℬ, are not the same, but considerably higher than for increasing M. M. F., and are

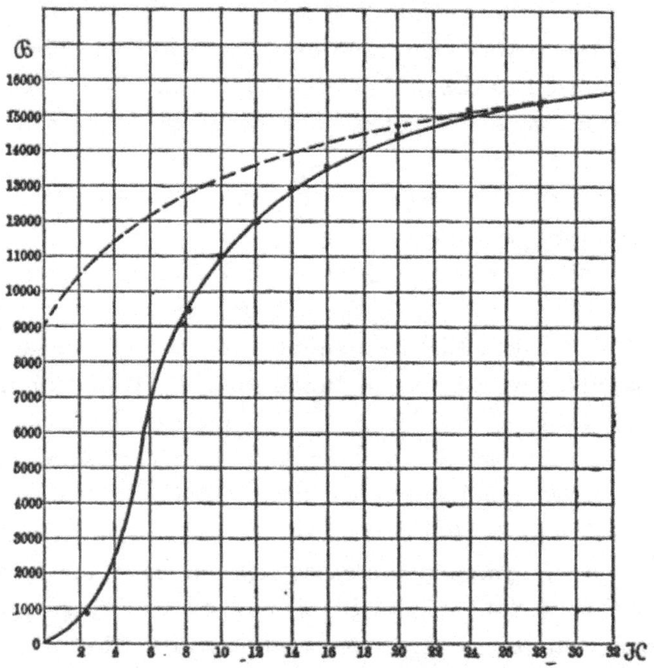

FIG. 29.—RISING AND DECREASING MAGNETIC CHARACTERISTIC.

shown in Fig. 29 in dotted line. Thus the magnetic characteristic is different for decreasing and for increasing magnetism ; or, in other words, the magnetic flux, ℬ, in magnetic materials, such as iron, depends not only upon the present value of M. M. F., but also upon the previous values, and thus lags behind the M. M. F.

This lag of the magnetic flux ℬ behind the M. M. F. is

practically independent of the time; that is, independent whether the change of M. M. F. takes place rapidly or very slowly.   It is called *hysteresis.*

The effect of hysteresis upon the magnetic characteristic is most pronounced in cyclic changes of flux as produced by cyclic changes of M. M. F. Thus if the M. M. F., $\mathfrak{F}$, or the

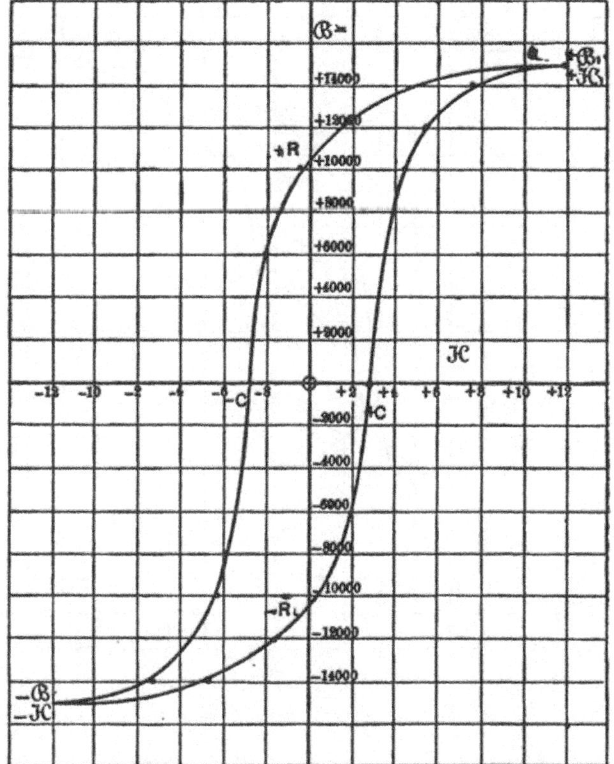

FIG. 30.—HYSTERETIC LOOP OR MAGNETIC CYCLE.

field intensity, $\mathfrak{K}$, is varied periodically between a maximum value $+\mathfrak{K}_1$ and an opposite value $-\mathfrak{K}_1$, the magnetic flux will vary between corresponding maximum values $+\mathfrak{B}_1$ and $-\mathfrak{B}_1$, describing a loop-shaped curve called the *magnetic cycle* or *hysteretic loop,* as shown in Fig. 30.

It follows that when the M. M. F. has been reduced to zero, the magnetic flux has still a considerable value, which is called the *remanent magnetism*, R in Fig. 30.

The magnetism will reach zero only after the M. M. F. has been reversed and increased to a considerable value $\mathfrak{C}$, in opposite direction. Thus the iron acts as if an internal

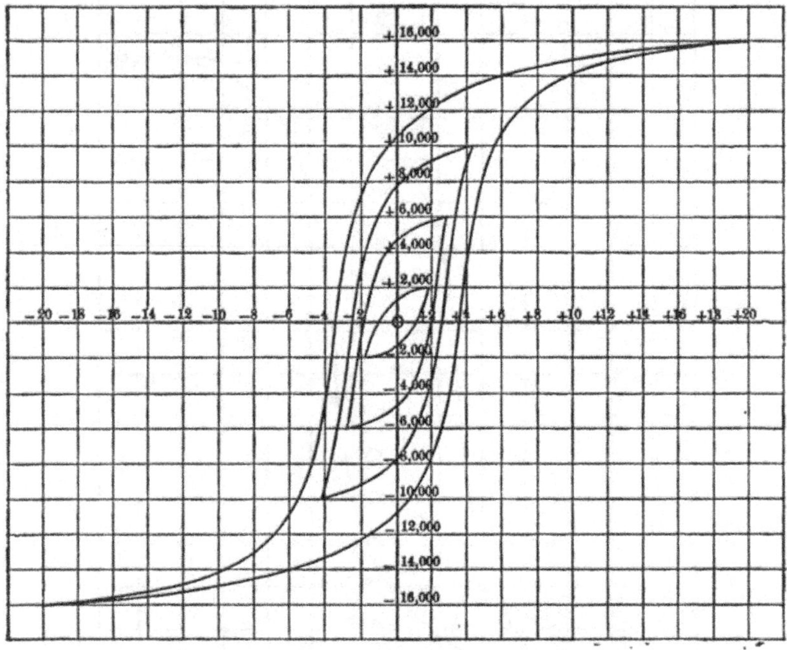

FIG. 31.—MAGNETIC CYCLES OF SOFT SHEET-IRON OR SHEET-STEEL.

M. M. F., $\mathfrak{C}$, tends to maintain its magnetic flux. This M. M. F. $\mathfrak{C}$ is called the *coercive force* of the iron.

The shape of the hysteretic cycles varies with different magnetic materials, and even with the same magnetic material in different physical conditions, and is different for different values of maximum magnetic flux, $\mathfrak{B}_1$. A number of such magnetic cycles are shown in Figs. 31, 32, and 33. Fig.

31 gives the magnetic cycles of sheet iron or soft sheet steel for different values of magnetic flux,

$$\mathfrak{B} = 2000, \quad 6000, \quad 10000, \quad \text{and} \quad 16000.*$$

Fig. 32 gives cast-iron cycles for $\mathfrak{B} = 6800$ and $\mathfrak{B} = 10300.$† Fig. 33 gives cycles of tool-steel at different degrees of hardness. H is a sample hardened in water, O one hardened in oil, and S an annealed sample, all three of the same

FIG. 32.—MAGNETIC CYCLES OF CAST-IRON.

material.‡ As may be seen, the harder the material the lower and wider in general is the hysteretic loop; that is, the lower the maximum and remanent flux, the higher the coer-

---

* "On the Law of Hysteresis," Part III, A. I. E. E. *Transactions*, 1894, p. 717.

† *Ibid.*, Part I, A. I. E. E. *Transactions*, 1892, p. 40.

‡ *Ibid.*, Part II, A. I. E. E. *Transactions*, 1892, p. 653.

cive force.   In the cycles of Figs. 31 to 33, the abscissæ are
not the field intensities $\mathcal{H}$, but the ampere-turns per centimetre length of the magnetic circuit, expressed by

$$F = \frac{10\,\mathcal{H}}{4\pi},$$

which is sometimes used as a practical unit of M. M. F.

If an air-gap is introduced into the magnetic circuit—that
is, if the magnetic circuit is partly of iron and partly of air,

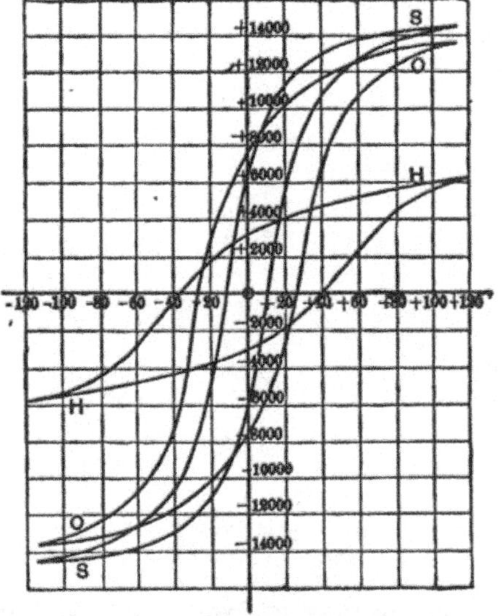

FIG. 33.—MAGNETIC CYCLES OF WELDED STEEL AT DIFFERENT DEGREES
OF HARDNESS.

as for instance in dynamo machinery—the hysteretic cycle
changes to the shape shown in Fig. 34, in which the straight
dotted line represents the M. M. F. required for the magnetization of the air-gap, and the hysteretic loop has the same
relative position—that is, the same horizontal distance from
this dotted line—as it had from the vertical line in the
circuit consisting entirely of iron.

As may be seen, the effect of the introduction of an air-gap is to require for the same maximum flux ℬ, a much greater M. M. F., to reduce the remanent magnetism very greatly, but the coercive force ℭ is not thereby affected.

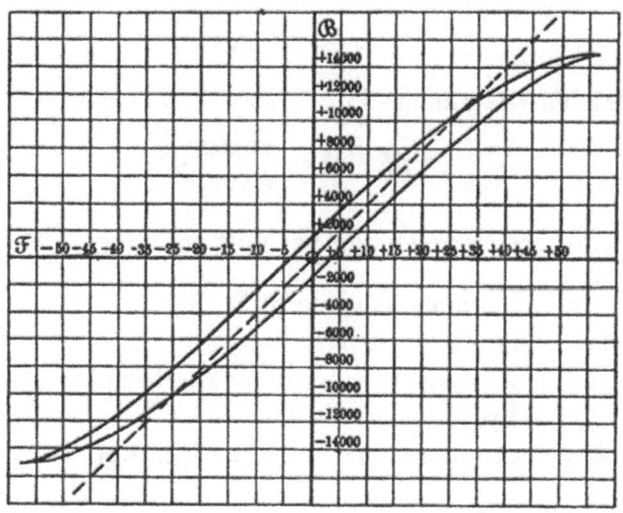

FIG. 34.—MAGNETIC CYCLE OF CIRCUIT CONTAINING AN AIR-GAP.

67. Hysteretic Loop.—In the hysteretic loop of a magnetic circuit, with M. M. F. as abscissæ, and magnetic flux as ordinates, the area of the loop is

$$\text{M. M. F.} \times \text{magnetic flux.}$$

Since M. M. F. has the dimension

$$L^{\frac{1}{2}} M^{\frac{1}{2}} T^{-1},$$

and magnetic flux the dimension

$$L^{\frac{3}{2}} M^{\frac{1}{2}} T^{-1},$$

where L = length, M = mass, T = time, the area has consequently the dimension

$$L^2 M T^{-2};$$

that is, the same as that of energy.

In the hysteretic loop with field intensity $\mathfrak{K}$ as abscissæ, of dimension

$$L^{-\frac{1}{2}}M^{\frac{1}{2}}T^{-1},$$

and magnetic flux density $\mathfrak{B}$ as ordinates, of dimension

$$L^{-\frac{1}{2}}M^{\frac{1}{2}}T^{-1},$$

the area has the dimension

$$L^{-1}M\,T^{-2} = \frac{\text{energy}}{\text{volume}},$$

thus representing an energy per unit volume.

Let $f =$ instantaneous value of M. M. F.,
  $\mathfrak{F} =$ maximum   "   "   "
  $\phi =$ instantaneous value of magnetism produced thereby,
  $\Phi =$ maximum value of magnetism produced thereby.

If the M. M. F. f is produced by an alternating current, i, flowing through n turns, then

$$e = -\,n\frac{\mathrm{d}\phi}{\mathrm{d}t} = \text{E. M. F.}$$

induced by the magnetism m, and

$$\mathrm{d}w = ei\mathrm{d}t = -\,ni\mathrm{d}\phi$$

$=$ energy expended by the change of magnetic conditions. Since $ni = f$,

$$\mathrm{d}w = -\,f\,\mathrm{d}\phi,$$

and

$$W = \int_{-\mathfrak{F}}^{+\mathfrak{F}} f\,\mathrm{d}\phi + \int_{+\mathfrak{F}}^{-\mathfrak{F}} f\,\mathrm{d}\phi =$$

total energy expended during the cyclic change of magnetism ; but

$$\int_{-\mathfrak{F}}^{+\mathfrak{F}} f\,d\phi + \int_{+\mathfrak{F}}^{-\mathfrak{F}} f\,d\phi = \text{area of the hysteretic loop.}$$

Therefore, the area of the hysteretic loop, with the M. M. F. in ampere-turns as abscissæ, and with the magnetic flux in volt-lines ($= 10^8$ lines, or one hundred mega-webers) as ordinates, is equal to the energy expended by hysteresis, in coulombs. With lines of force as ordinates, and tens of ampere-turns as abscissæ, the area is the hysteretic energy in ergs.

The area of the hysteretic loop, with field intensity, $\mathcal{H}$, in tens of ampere-turns per unit length of magnetic circuit, as abscissæ, and with magnetic flux density, $\mathcal{B}$, as ordinates, is equal to the loss of energy by hysteresis in ergs per unit volume. With field intensity $\mathcal{H} = \dfrac{4\pi\mathfrak{F}}{10}$ as abscissæ, and with lines of force per cm.', $\mathcal{B}$, as ordinates, the energy expended by hysteresis during a complete cycle of magnetization is $= 4\pi \times$ area of hysteretic loop.

Hysteresis thus represents an expenditure of energy by the M. M. F. and is measured by the area of the hysteretic loop or magnetic cycle.

68. Molecular Magnetic Friction.—If by an alternating M. M. F. an alternating magnetic flux is produced in iron or other magnetic material, a loss of energy takes place in the iron by a kind of frictional resistance of the molecules against the change of their magnetic condition. This phenomenon is called *molecular magnetic friction.* Therefore, to alternate a magnetic flux, energy has to be expended upon the iron.

If the alternation of the magnetic flux is produced by an alternating current, and the condition is such that no energy is expended upon the magnetic circuit by any other source, nor external work done by the magnetic circuit, the energy consumed by molecular magnetic friction has to be supplied by the alternating current. Consequently, the magnetic flux cannot follow the M. M. F., but must lag behind it so far that the hysteretic curve of magnetic flux and M. M. F. represents the energy expended by molecular magnetic friction.

It follows that in an alternating magnetic circuit which neither produces external work nor receives energy from another source than the alternating M. M. F., the energy consumed by molecular magnetic friction is equal to the energy expended by magnetic hysteresis.

If, however, external work is done by the magnetic circuit, or work expended upon it by an external force, the identity between the energy of molecular magnetic friction and the energy of magnetic hysteresis no longer exists. Thus, if the magnetic circuit is vibrated mechanically during the cycle of magnetization, the hysteretic loop collapses more or less completely, and the rising and decreasing magnetic characteristics coincide ; the energy consumed by molecular magnetic friction being supplied in this case from the mechanical source vibrating the magnetic circuit. Conversely, if mechanical work is done by the magnetic circuit, as, for instance, if the magnetic circuit consists of iron filings or loose laminations which can vibrate and rearrange themselves, the hysteretic loop is greatly extended and represents not only the energy consumed by molecular magnetic friction, but also the mechanical work done.*

---

* For proof and discussion of the distinction between hysteresis and molecular magnetic friction see: ''On the Law of Hysteresis,'' Part II, Chap. V, A. I. E. E. *Transactions*, 1892, p. 711, and ''On the Law of Hysteresis.'' Part III, Chap. II, A. I. E. E. *Transactions*, 1894, p. 706.

It follows that in determining the energy loss by molecular magnetic friction from the hysteretic loop of the material, care must be taken that neither external work is done nor absorbed by the magnetic circuit while the hysteretic loop is being determined.

69. Determination of Hysteresis and Molecular Magnetic Friction.—The different methods of determining the value of hysteresis and molecular friction are as follows:

(*a*) *Ballistic Method.*—A magnetic circuit is built up of the iron to be tested, a magnetizing coil wound around it as uniformly as possible, and a second or exploring coil employed, connected to a ballistic galvanometer. The current in the magnetizing coil is varied step by step, and the time integral of E. M. F. induced in the exploring coil by the variation of current, and consequently the change in magnetic flux, is observed by means of the ballistic galvanometer. In this way a complete cycle of magnetism is plotted and from its area the loss of energy determined. This method does not give the energy of molecular friction directly, but gives the energy expended by hysteresis. It can be used for the determination of the saturation curve also, and is suitable for the investigation of solid materials, as well as of laminations, etc. In determining the saturation curve by this method, it is desirable to dissipate the remanent magnetism previous to the test, by applying a strong alternating current through the magnetizing coil, and gradually reducing this current to zero. In determining the hysteretic cycle, a greater number of cycles between the same maximum values should be described before taking readings, so as to make the cycle symmetrical and independent of remanent magnetism due to the previous history of the iron.

(*b*) *Alternate Current Method.*—The energy expended by magnetic hysteresis can be determined directly by sending an alternating current through a magnetizing coil surround-ing the magnetic circuit, and taking readings of an ammeter, a voltmeter, and a wattmeter. The M. M. F. is determined from the ammeter reading, the magnetic flux by calculation from the voltmeter reading, number of turns, frequency and shape of the magnetic circuit, and the energy loss by hysteresis from wattmeter readings.

This method is applicable only to laminated material, and measures not only the energy expended by hysteresis but also the energy loss by eddy or Foucault currents. Consequently, either the material has to be subdivided so as to make the latter negligible, or hysteresis and eddy currents have to be separated from each other afterwards.

Since the maximum flux and maximum M. M. F. in this method are obtained by calculation from the effective values read on the instruments, the wave of E. M. F. should be as near as possible a sine wave. Owing to the distortion of the current wave by hysteresis, the maximum value of current, even with a sine wave of E. M. F., is not throughout the whole range equal to $\sqrt{2}$ times the effective value. Calculating the M. M. F. under the assumption of a sine wave of current, therefore, gives a magnetic characteristic, which, while practically coinciding with the true magnetic characteristic within a range up to $\mathfrak{B} = 10{,}000$ or $14{,}000$, yet differs greatly therefrom beyond this range, showing apparently very high values of flux. In Fig. 35 is shown the true magnetic characteristic in full line, the magnetic characteristic as determined from an alternating current test in dotted line, and the loss of energy by hysteresis in ergs per cm³ and cycle in the full line of single curvature.*

Instead of connecting the voltmeter and the potential coil of the wattmeter across the magnetizing coil of the

---

* "On the Law of Hysteresis," Part III, A. I. E. E. *Transactions*, 1894, p. 722.

magnetic circuit, it is preferable to connect it across a
second or exploring coil wound uniformly over the magnetic

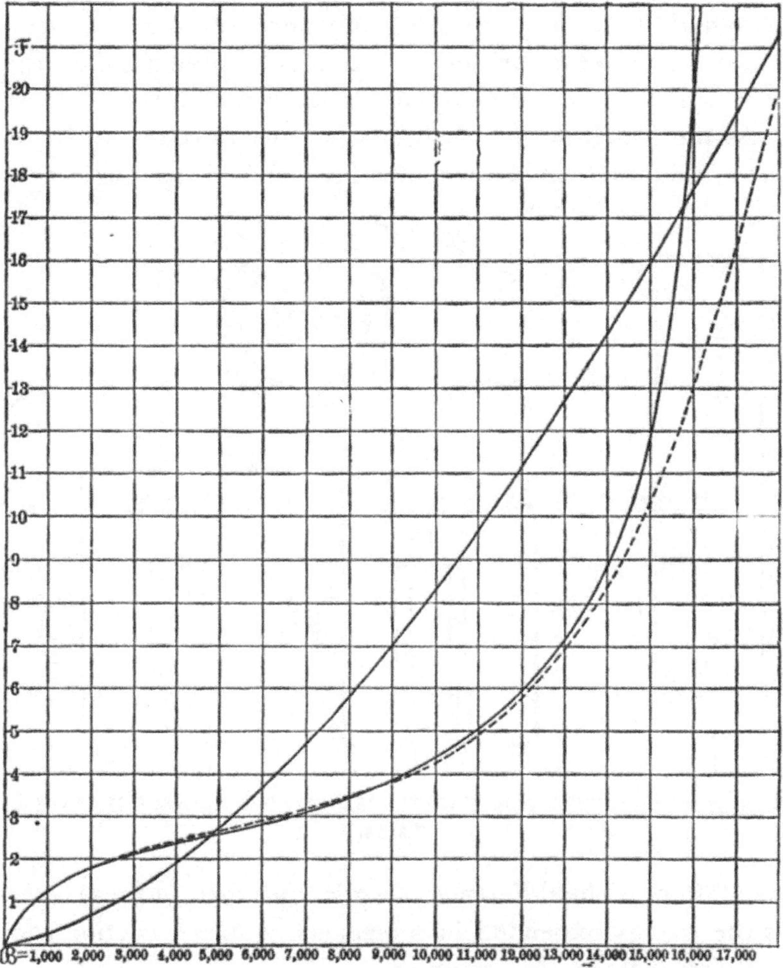

FIG. 35.—MAGNETIC CHARACTERISTIC FROM BALLISTIC AND ALTERNATING
CURRENT TESTS, AND MOLECULAR MAGNETIC FRICTION CURVE.

circuit, as shown diagramatically in Fig. 36, and thereby
eliminate the error due to the drop of voltage and loss of
energy in the resistance of the magnetizing coil.

(*c*) *Power and Torque Tests.*—The loss of energy by mo‑lecular magnetic friction and eddy currents may be deter‑mined by moving the material to be tested in a uniform magnetic field, and measuring the power required therefor.

This method is commonly used in determining the hyster‑etic loss in the armatures of dynamo machinery. In this case the armature is turned, first in an unexcited magnetic field, and then in a magnetic field of various degrees of

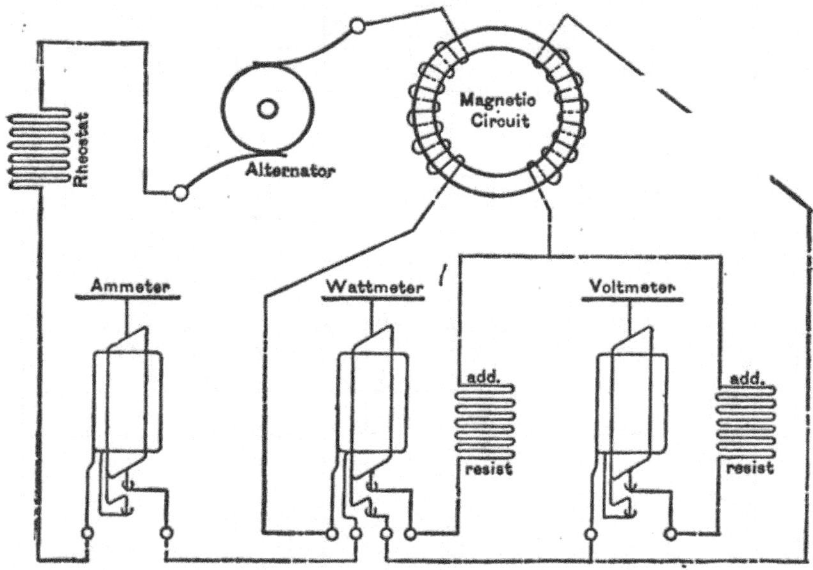

FIG. 36.—INSTRUMENT CONNECTIONS FOR ALTERING CURRENT HYSTERESIS TESTS.

excitation. The difference in work consumed in these cases is the energy expended in molecular magnetic friction and eddy currents in the armature.

A similar method suitable for testing of iron, is the following :

A uniform magnetic field is set in rotation, and in the centre of the field the iron to be tested is suspended by a torsion spring. Owing to the hysteretic loss in the iron, the rotating field tends to turn the iron in its direction of rotation, and the torsion given to the spring to counterbalance this torque is directly proportional to the hysteretic loss per cycle in the test-piece. The magnetic flux is measured by the E. M. F. induced in an exploring coil surrounding the test-piece.* This method has the advantage that the torque exerted upon the iron by molecular magnetic friction is independent of the speed of the rotating field.

70. Loss of Energy.—The hysteretic cycle has been found identically the same from very slow alternations, up to the highest frequencies reached by dynamo-electric machinery—beyond 200 cycles per second. It can, therefore, be said that hysteresis and the loss of energy by molecular magnetic friction per cycle are independent of the frequency, and consequently the loss of power is proportional to the frequency, that is, to the number of cycles.

No difference between the value of hysteretic energy and loss by molecular magnetic friction has been found between rotating and reversing fields, and therefore it is permissible to use the values found in alternating fields for losses in rotating fields, and conversely.

The hysteretic loop increases with increasing maximum magnetic flux density $\mathfrak{B}$; that is, the energy expended by hysteresis per cycle, and the energy absorbed by molecular magnetic friction, both increase with the magnetic flux. Plotting the loss of energy as a function of the magnetic flux, the curve thus obtained rises uniformly, and does not show any marked features at the critical points of the magnetic saturation curve, following the same law at saturation and below saturation.

Approximately, the loss of energy by molecular magnetic

---

* Holden, *The Electrical World*, June 15, 1895.

friction can be expressed by the empirical formula,*

$$W_H = \eta \, \mathfrak{B}^{1.6},$$

where $W_H =$ loss of energy in ergs per cycle and cm^3, $\pm\mathfrak{B} =$ maximum values of magnetic flux between which the magnetic cycle is performed, and η is the *coefficient of molecular magnetic friction.*

The same empirical function of the 1.6 power holding for reversals of magnetism, holds also for cyclic changes of

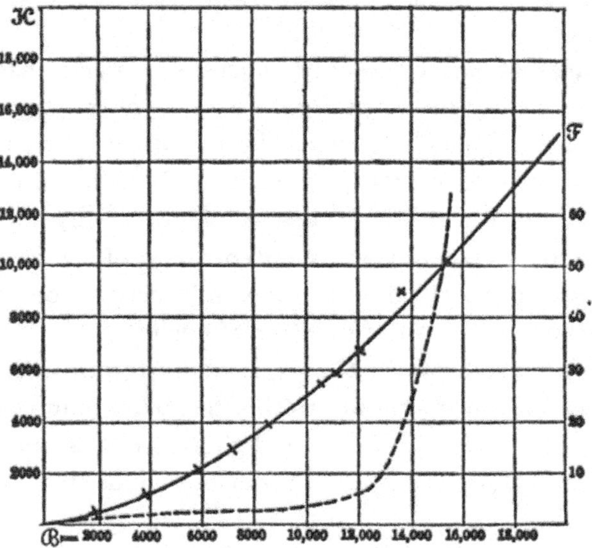

FIG. 37.—CURVE OF HYSTERESIS OF SOFT IRON WIRE.

flux between any two limits $\mathfrak{B}_1$ and $\mathfrak{B}_2$, whether these two limits be of the same direction and same sign or of opposite direction. Thus, in general form the empirical law of molecular magnetic friction can be expressed by the formula :

$$W_H = \eta \left(\frac{\mathfrak{B}_2 - \mathfrak{B}_1}{2} \right)^{1.6}.$$

---

* "On the Law of Hysteresis," A. I. E. E. *Transactions*, 1892, p. 3 and p. 621; 1894, p. 702

This equation can be considered as empirical only, and, in fact, does not hold for extremely low values of magnetic flux; it is, however, correct with sufficient approximation through a wide range, as shown in Figs. 37 and 38,* which give the observed values of hysteretic loss in a sample of soft-iron wire, and in a sample of annealed steel wire, as taken from Ewing's tests.　The curve of 1.6 power is shown in full

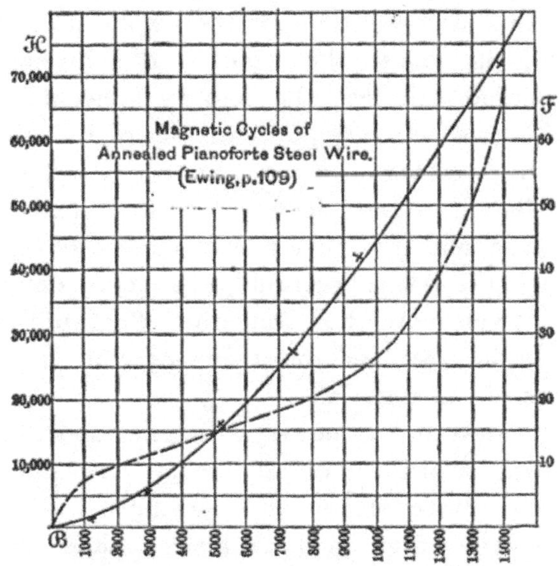

FIG. 38.—CURVE OF HYSTERESIS OF ANNEALED PIANOFORTE STEEL WIRE.

line in these figures, with the observed values marked by crosses.

The dotted line represents the magnetic characteristic.

In reality, the value $\mathfrak{B}$ in the formula of hysteresis should probably not be the total magnetic flux, but the "metallic magnetic flux," or $\mathfrak{B} - \mathfrak{H}$, which reaches a finite value of

---

* "On the Law of Hysteresis," Part II, A. I. E. E. *Transactions*, 1892, pp. 673, 674

absolute saturation, while $\mathfrak{B}$ increases indefinitely with increasing $\mathfrak{H}$.*

71. Coefficient of Hysteresis.—The coefficient of molecular magnetic friction varies very greatly with different materials and even with different conditions of the same material.

In iron, the loss by molecular magnetic friction seems to depend comparatively little upon the chemical constitution. In general, the purer the iron is, the lower the coefficient η, and the less the loss by molecular friction. Frequently, however, very pure samples of iron show comparatively high values of η, while impure samples show remarkably low hysteretic losses. A large percentage of carbon, silicon, and phosphorus seems to be objectionable, while manganese in small percentages appears comparatively harmless. Even here, however, the effect of impurities seems to be indirect, and due to changes in the physical constitution of the material.

Of the greatest importance regarding hysteretic loss is the *physical condition* of the material, one and the same material in a hardened state occasionally showing a hysteretic loss many times larger than when in annealed state. Annealing is always found to reduce the loss by molecular friction more or less, while hardening increases it. With tool steel, for instance, the coefficient of hysteresis has been varied from 14.5 to 75 by annealing and hardening, respectively. (See Fig. 33.)

Break of continuity of the material usually greatly increases the loss by molecular magnetic friction. Thus, gray cast-iron, even when very soft, shows a comparatively high

---

* "On the Law of Hysteresis," Part II, A. I. E. E. *Transactions*, 1892, Chap. III, pp. 678 *seq.*

coefficient, η, due to the interposition of graphite in the metallic structure.

The *temperature* affects the loss of energy by molecular magnetic friction very little within the range of atmospheric temperatures, the hysteretic loss decreasing considerably only at high temperatures. If iron is heated and then cooled, the loss by molecular friction is much smaller when hot, and is not increased again to the same value as before when cooled ; a part of the decrease due to the heating becomes permanent, probably due to a change of physical condition. This is especially noticeable with steel. After repeated heating and cooling, the variation becomes reversible and the hysteretic loss decreases with increasing temperature and increases again to the same value with decreasing temperature, approximately as a linear function of the temperature.*

Mechanical action affects the energy expended by hysteresis very greatly, and causes the hysteretic loop to collapse more or less, but apparently does not affect the loss of energy by molecular magnetic friction; a distinction thus exists between hysteretic loss and molecular magnetic friction loss.

Very long-continued exposure of iron in alternating magnetic fields seems in some cases to increase the loss by molecular magnetic friction through what has been called *ageing* of the iron, which has been observed especially in iron with very low coefficient η. Other careful tests, however, have not shown any trace of an increase of hysteretic loss during continuous use in alternating fields, and the observed increase thus appears to be due to secondary causes, probably to a continued heating in the alternating field beyond a critical point of the iron, and a change of the

---

* W. Kunz, *Electrotechnische Zeitschrift*, Berlin, April 5, 1894.

physical condition caused thereby. An effect similar to that observed in the crystallization of wrought-iron in bridges, etc., under vibrating stresses, may also account for the so-called ageing.

The following values of the coefficient of molecular magnetic friction have been observed in different materials.[*]

COEFFICIENT OF HYSTERESIS, AND ABSOLUTE MAGNETIC SATURATION.

	Coefficient in Milliunits.	Average	Absolute Magnetic Saturation, $\mathcal{B} - \mathcal{H} = 4\pi I$ in Kilolines.
Soft sheet-iron and sheet-steel............	1.24–5.5	2.5–3.5	17–20
Cast-iron............................	11.3–16.2	13	10–11
Cast-steel of low permeability..........	12	......	11
" of high " soft.....	3.32–9	6	12–19.5
" of " " hard....	28	....	18.5
Welded steel annealed............. ...	14.5	....	17.4
" " oil hardened.............	27	....	16.7
" " very hard................	75	....	8.3
Manganese steel, annealed, 4.7% Mn...	41		
" " " 8.74% Mn..	82		
" " oil hardened, 4.7% Mn	67		
Chrome-steel, annealed, 1.2% Cr.......	16		
" " oil hardened, 1.2% Cr....	44		
Wolfram steel, annealed, 4 6% Wo.....	14		
" " oil hardened, 3.44% Wo.	48		
Magnetite............................	20.4–23.5	....	4.7
Nickel wire, soft.....................	12.2–15.6	13	5.9
" " hardened................	38.5		
Cobalt, cast.........................	11.9		
Amalgam of iron, 11% Fe.....	231	....	.9

While the loss of energy by hysteresis and by molecular magnetic friction is independent of the frequency within a very wide range, for extremely slow variations of M. M. F. a time lag exists, especially for very low M. M. F.'s. That

---

[*] "On the Law of Hysteresis," Part II, A. I. E. E. *Transactions*, 1892. p. 680; Part III. 1894, p. 705.

is, after the application of the M. M. F. the magnetism rises quickly to a certain value and then keeps on rising slowly for seconds and even minutes until a final second value is reached. In alternating current machinery this phenomenon of magnetic sluggishness is of no importance.

72. Eddy Currents.—In magnetic materials exposed to an alternating magnetic field, besides the loss of energy by magnetic hysteresis or molecular magnetic friction, a further loss of energy takes place by *eddy currents or Foucault currents.* The eddy currents are not a magnetic phenomenon like hysteresis, but a purely electrical phenomenon. They are secondary currents induced in the iron by the alternating magnetic field. Thus, while the loss of energy by molecular magnetic friction is entirely independent of the shape of the magnetic circuit, and is the same whether the material is solid or subdivided, the loss of energy by eddy currents depends entirely upon the shape of the iron, and though often excessive in solid iron, may be made small or negligible by thorough subdivision.

Eddy currents are true secondary or induced currents, and therefore their E. M. F. is proportional to the magnetic flux $\mathcal{B}$ and the frequency N; consequently the loss of energy by eddy currents is proportional to $\mathcal{B}^2$ and N^2. The loss of energy per cycle is proportional to the square of the magnetic flux, $\mathcal{B}$, and directly proportional to the frequency N, while the loss of energy of molecular magnetic friction is proportional to the 1.6th power of the magnetic induction $\mathcal{B}$ and independent of the frequency N.

The loss of energy by eddy currents per cm^3 and cycle may, therefore, be represented by the formula

$$W_e = \epsilon N \mathcal{B}^2 \, ;$$

and the total loss of energy in the iron by molecular magnetic friction and eddy currents by the formula

$$W = W_H + W_e = \eta \mathfrak{B}^{1.6} + \epsilon N \mathfrak{B}^2,$$

where $\eta =$ coefficient of hysteresis;
   $\epsilon =$ coefficient of eddy currents;
   $N =$ frequency;
   $\mathfrak{B} =$ maximum magnetic flux density;
   $W = W_H + W_e =$ total loss of energy, in ergs per cm³ and cycle.*

This formula of the total loss of energy allows the separation of the loss by molecular magnetic friction from the loss by eddy currents, by means of two experimental observations. This can be done either by making two observations for the same flux $\mathfrak{B}$ and different frequencies N_1 and N_2, or by making two observations at the same frequency N for different magnetic fluxes $\mathfrak{B}_1$ and $\mathfrak{B}_2$. In either case we get two equations with two unknown quantities η and ϵ. The latter method, with different magnetic fluxes and the same frequency, is very convenient, but depends upon the shape of the curve of molecular magnetic friction—that is, upon the empiric law of the 1.6th power—while the former method of observation at different frequencies is independent thereof.

In general, it is preferable to make a considerable number of observations, plot them in a curve, and take two points from the curve as far apart from each other as is possible without introducing the increased error of observation at the limits of the range of test.

---

* "On the Law of Hysteresis," Part I, A. I. E. E. *Transactions*, 1892, p. 22. For calculation of the coefficient of eddy currents for laminations and wire, see "On the Law of Hysteresis," Part III, A. I. E. E. *Transactions*. 1894, pp. 734–738.

73. Effect of Molecular Magnetic Friction and Eddy Currents.—Eddy currents can be eliminated by thorough subdivision of the magnetic material, and therefore no excuse exists for their presence in any first-class electrical machinery.

Molecular magnetic friction represents a loss of energy which cannot be avoided by subdivision, but can be greatly reduced by the choice of the best available iron. This loss of energy is present wherever magnetic fields are alternating, and is, therefore, found in the armatures of continuous current dynamos and motors; to a larger extent still in the armatures of alternators, due to the higher frequency at which they usually operate; and in transformers and induction motors.

In continuous current machinery the loss by molecular magnetic friction is usually only a comparatively small part of the total loss. In alternators the molecular magnetic friction loss is one of the largest losses, and frequently even larger than all the other losses combined. It is therefore of the greatest importance in alternating practice to use the best possible iron. It is obvious that neglect of the molecular magnetic friction in the determination of the efficiency of alternators causes the efficiencies to appear very much higher than they really are. Consequently the so-called electrical efficiency of alternators is of no value whatever, and in discussing the efficiency of alternators, care has to be taken to include the loss of molecular magnetic friction; this has not always been done, and as a result efficiencies have been, and still are being, quoted in excess of any reasonable value. When speaking of alternators, obviously alternate current generators as well as synchronous motors are understood.

In transformers the loss from molecular magnetic friction is usually the largest of the several losses, especially in larger sizes of transformers, and is the more undesirable as it occurs equally at all loads and at no load, while the resistance loss in the electric conductors is significant only with a considerable load. Thus the all-day efficiency of a transformer, or the ratio of the total amount of energy put into the primary of the transformer to the useful energy taken out at its secondary, depends almost exclusively upon the molecular magnetic friction loss. The same applies to induction motors, in so far as they are transformers.

74. Equivalent Sine Curves.—A further effect of molecular magnetic friction is its action on the shape of the current wave. If a sine wave of E. M. F. is impressed upon a magnetic circuit, a current will be produced in the circuit which is not a sine wave, but differs widely therefrom, being distorted by hysteresis.

With a negligible internal resistance, the sine wave of impressed E. M. F. implies a sine wave of counter E. M. F. or induced E. M. F.—that is, a sine wave of magnetism, or of magnetic flux, $\mathfrak{B}$. Owing to the discrepancy between M. M. F. and magnetic flux as represented in the hysteretic cycle, the M. M. F., and therefore the current required to produce the sine wave of magnetic flux, is not a sine wave. It is determined by taking from the sine wave of magnetic flux $\mathfrak{B}$ the instantaneous values of $\mathfrak{B}$, and from the hysteretic curve the values of M. M. F.; and therefore of current corresponding to the instantaneous values of magnetic flux $\mathfrak{B}$, and plotting these currents as functions of the time in the same way as magnetic fluxes are plotted as sine waves. In this way different current waves are produced for different maximum magnetic fluxes. A number of such distorted current waves are shown in Figs. 39 to 42, corresponding to values of maximum flux

$$\mathfrak{B} = 2000, \quad 6000, \quad 10{,}000, \quad 16{,}000$$

and of M. M. F.'s,

$$\mathfrak{F} = \quad 1.8, \quad\quad 2.8, \quad\quad 4.3, \quad\quad 20.0,$$

and also to the hysteretic cycles shown in Fig. 31. As may be seen, all these curves are bulged out at the ascending,

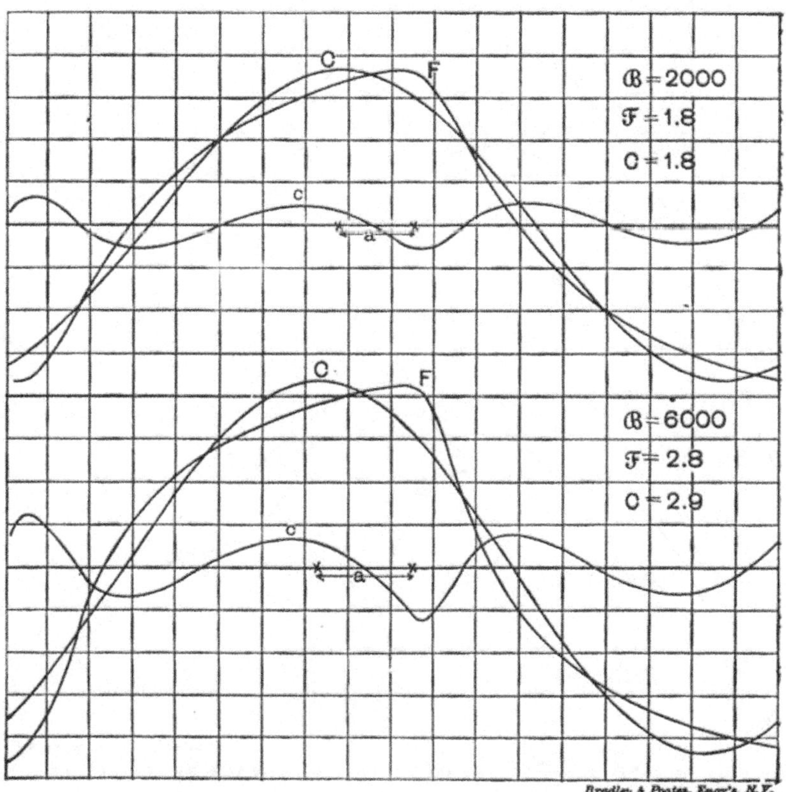

FIGS. 39 AND 40.—DISTORTION OF CURRENT WAVE BY HYSTERESIS.

and hollowed in at the descending side. With increasing values of maximum magnetic flux, the maximum point of the current wave becomes pointed, as in Fig. 41, and ultimately a sharp high peak is formed, as in Fig. 42, due to magnetic saturation.* Such distorted current waves can, like most distorted alternating waves, be replaced for all practical purposes by *equivalent sine waves ;* that is, *sine waves equal in effective intensity and in energy* to the distorted wave. These equivalent sine waves of exciting current are shown

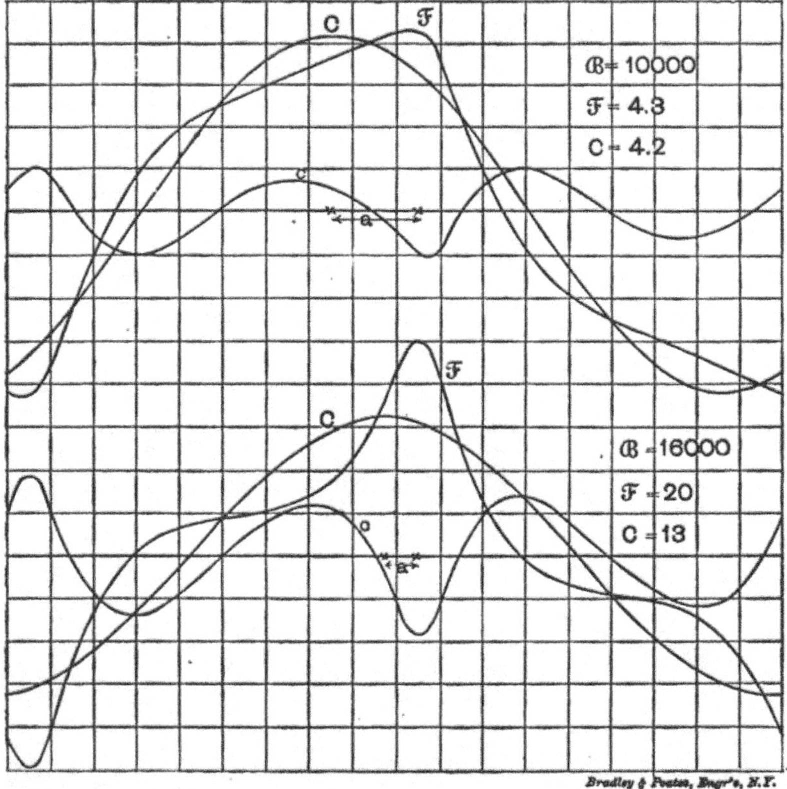

FIGS. 41 AND 42.—DISTORTION OF CURRENT WAVE BY HYSTERESIS.

in Figs. 39 to 42 ; the remainder, or the difference between the true distorted current wave and the equivalent sine wave, is shown also, and consists, as seen, essentially of a term of triple frequency. This means that hysteresis intro-

---

* "On the Law of Hysteresis," Part III, A. I. E. E. *Transactions,* 1894, p. 719.

duces higher harmonics in the current wave, amongst which the triple harmonic is especially pronounced.

The equivalent sine wave of exciting current is not in phase with the wave of magnetism, but leads it by an angle α, which is called the angle of hysteretic advance of phase. In consequence, the *exciting current* can be resolved into two components; one,

$$I \cos \alpha,$$

in phase with the magnetism—that is, in quadrature with the induced E. M. F., or wattless, which is called the *magnetizing current*; and into another component,

$$I \sin \alpha$$

in quadrature with the magnetism—that is, in phase with the induced E. M. F., representing consumption of energy and called the *magnetic energy current*.

The ratio of the magnetic energy current to the total exciting current is the so-called "power" factor of the electromagnetic circuit, whose value is

$$\sin \alpha = \frac{4\eta\mu}{\mathfrak{B}^{0.4}};*$$

that is, it depends upon the coefficient of hysteresis η, the magnetic permeability μ and the maximum magnetic flux $\mathfrak{B}$, but is entirely independent of the shape of the magnetic circuit and the volume of iron used; in other words, it is a function of the iron alone and not of the electric circuit.

Eddy currents have in general no effect on the shape of the wave, but follow the shape of the E. M. F. wave like any other secondary current. In their influence upon the pri-

---

* "On the Law of Hysteresis," Part III, A. I. E. E. *Transactions*, 1894, p. 595.

mary current they also act like a secondary current—that is, in the transformer like a partial load on the secondary circuit, shifting the current wave ahead of the magnetism, and bringing it into lesser phase displacement from the **E. M. F.**

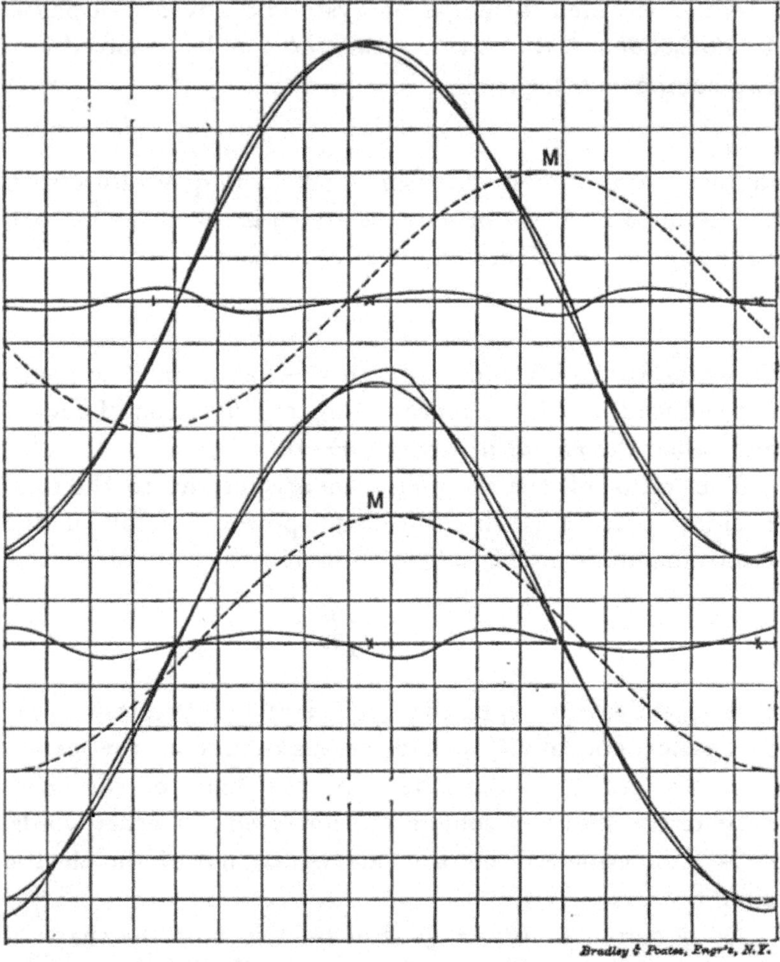

FIGS. 43 AND 44.—DISTORTION OF CURRENT WAVE BY HYSTERESIS.

In Fig. 43 is shown the current wave in a transformer at partial secondary load, showing the reduction of the distortion under load. The dotted line *M* represents the magnetic flux.

In a magnetic circuit consisting partly of iron and partly of air, the curve of exciting current of an impressed sine wave of E.M.F. is the sum of the exciting currents of the iron and of the air portions of the circuit. Since air has no hysteresis, but very low permeability, the exciting current of the air portion of the circuit is a sine wave of high amplitude, and therefore, when superimposed upon the wave of exciting current of the iron circuit, reduces the relative distortion of the total exciting current, and makes it appear more sine-shaped ; this is shown in Fig. 44, which represents the exciting current of a magnetic circuit containing an air-gap of 1/400 of the length of the iron,* and corresponding to a maximum flux of $\mathfrak{B} = 6000$, or to the curve shown in Fig. 40. The magnetic flux is shown by the dotted line M.

75. Hysteretic Losses.—The loss of energy by hysteresis depends upon the maximum value of the magnetic flux, while the energy transformed in a transformer depends upon the effective value of E. M. F. With a sine wave of E. M. F., the magnetic flux follows a sine wave also, and therefore the maximum value of the latter is proportional to the impressed E. M. F. If, however, the wave of impressed E. M. F. is distorted, or differs from the sine form, the wave of magnetic flux will be distorted also by the superposition of higher harmonics, and therefore, with the same effective value of impressed E. M. F., its maximum value in the case of a flat-topped wave of magnetic flux, can be lower than with a sine wave, or higher than with a sine wave with a peaked wave, of magnetic flux. In the first case the same amount of energy will be transformed with a lesser hysteretic loss, and in the latter case with a larger hysteretic loss; or, in other

---

* "On the Law of Hysteresis," Part III, A. I. E. E. *Transactions*, 1894, p. 720.

words, the hysteretic loss in a transformer, at the same impressed effective E. M. F., depends upon the shape of the E. M. F. wave, and differs with different wave-shapes of E. M. F., by as much as 30%.

Iron-clad alternators of "uni-tooth" construction in general give peaked waves of E. M. F., and, therefore, flat-topped waves of magnetic flux ; that is, give hysteretic losses in transformers lower than with sine waves by something like 10%.

Distributed windings, that is, comparable to windings of a continuous current machine tapped at two diametrically opposite points of the armature, frequently give a pointed wave of magnetic flux, thus increasing hysteretic losses.

Since eddy currents are secondary currents, the loss of energy by such currents in transformers is entirely independent of the wave-shape of impressed E. M. F., and therefore proportional to the load on the secondary circuit at constant resistance ; or, in other words, if the secondary circuit of the transformer is closed by a given resistance, and the same effective E. M. F. impressed upon the primary circuit, the energy of this E. M. F. will divide in the same proportion between the useful secondary circuit, and the secondary waste circuit of eddy currents, independently of the shape of the E. M. F. wave.

ELECTRICITY.

PROPERTIES OF ELECTRIFIED BODIES.

76. Phenomenon of Electrification.—When a glass or resin rod is rubbed with a piece of woollen cloth, it is observed that the rod attracts light bodies. The experiment may be made by suspending a ball of elder-pith by a silk thread. The ball is first attracted by the rod, then it is repelled after having touched the rod. But if a rod of rubbed resin is brought near the ball, after it has been repelled by the glass rod, it is observed that the ball is again attracted.

The bodies between which similar actions are shown are said to be *electrified*, and the name *electricity* is given to the unknown agent which produces these phenomena.

Observations hows that the electric properties observed in glass and resin are general. Any two bodies, A and B, attract each other after having been rubbed together. But two bodies of the same nature, A and A', repel each other after having been rubbed by a third.

To interpret these phenomena, opposite electrifications are attributed the bodies after rubbing; those which behave like glass, in regard to wool, are said to be positively electrified, or charged with *positive electricity;* those which act like resin are said to be negatively electrified, or charged with *negative electricity*. Electric actions are summed up in the following rule:

Bodies charged with electricity of the same name repel each other and attract those charged with electricity of contrary name.

It must be observed, however, that these denominations do not imply the existence of two distinct kinds of electricity, but that they are only forms of speech designed to denote different states of electrification. One must not extend their meaning further than is done in mathematics in the case of quantities of contrary sign.

Following a hypothesis suggested by Franklin, electricity is generally compared to an imponderable fluid, of which every body contains a normal quantity. If the charge surpasses this quantity, there is positive electrification. In the opposite case the electrification is negative. A body is in the neutral state when it has its normal quantity of electricity. According to some eminent physicists, Clausius among others, electricity is supposed to be the ether in which the molecules of all bodies are enveloped and which fills interplanetary space. The phenomena of electrification are, on this hypothesis, supposed to be due to particular conditions of the ether.

77. Conductors and Insulators.—Different bodies act differently with regard to electricity. When one of the extremities of a metallic rod is electrified by friction, attractive properties are immediately manifested in all parts of the rod. In a rod of resin, on the contrary, this propagation of the phenomenon is extremely slow. Those bodies which propagate electric actions rapidly are called *conductors*, the rest are called *insulators*.

Conductors, which include all metals and alloys, have to be supported by insulators in order to retain their attractive properties. From the point of view of the experiments which we are examining, the earth and all bodies impregnated or coated with moisture act like conductors. The choice of the insulating supports is therefore important for the success of these experiments. Air is frequently employed as an insulator ; in order to prevent the moisture it contains from condensing on the solid insulators which support the electrified conductors, and thus producing a moist conductive coating, a support may be used whose base passes through air dried by concentrated sulphuric acid, Fig. 45.

FIG. 45.

Among solids the insulators most commonly used are glass, glazed porcelain, india-rubber, ebonite or india-rubber hardened by combination with sulphur, gutta-percha, paraffine, silk, cellulose, and shellac. Some of these substances, such as glass and cellulose, are hygroscopic. They should be covered with a coating of insulating varnish or impregnated with paraffine.

78. Electrification by Influence.—The experiment with the pith-ball, § 76, has shown us that a body can be electrified by contact. Even the mere approach of an electrified body is sufficient to produce electrical manifestations in a neighboring conductor.

Thus a sphere, A, charged with positive electricity, being brought over a conductor BC, Fig. 46, it is observed that the extremities of this latter act upon a freely suspended pith-ball. If this ball be charged beforehand with positive electricity, it is seen to be attracted by the extremity B, and repelled by C. From this we conclude that the former is negatively, the latter positively, electrified. When the influencing sphere, A, is withdrawn, the conductor, BC, no longer acts on a ball in the neutral state ; which shows that

the contrary electricities accumulated in B and C have
re-established the neutral state by recombining.

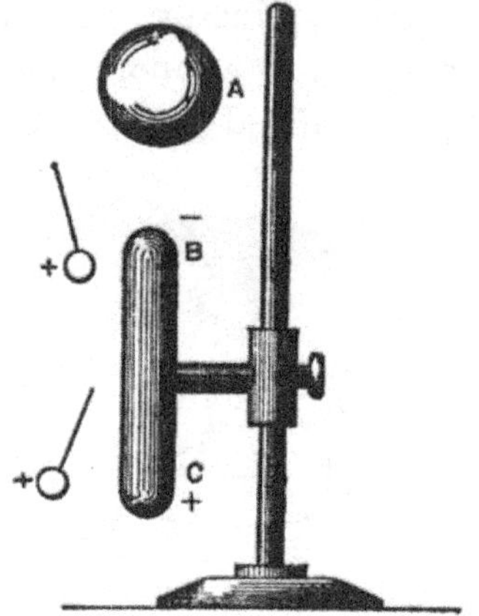

FIG. 46.

If the conductor BC is connected for a moment to the
earth while it is under the influence of A, we observe after
taking away A that every portion of BC is charged
negatively.

This method of charging the conductor is called electrification by influence.

79. Let us suppose that the framework of a quadrant
electrometer, E, Fig. 47, be connected to the earth through

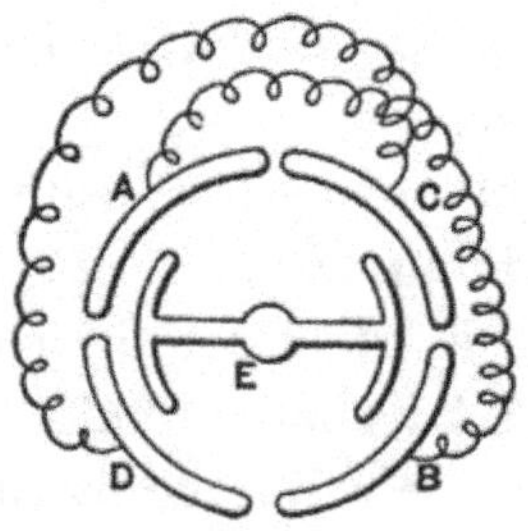

FIG. 47.

the suspension-wire, and that the two pair of quadrants are
electrified, but of opposite sign. The framework will be
electrified by influence; it will be equally attracted in both

directions and will remain motionless. But if it receives a positive charge, it will immediately be displaced towards the negative quadrants. If the charge be negative, the displacement will be towards the positive quadrants. We shall show, further on, that the swing of the framework is proportional to the charge that has been given to it; we shall also see that in order to give the quadrants equal and contrary charges they need only be connected to the poles of a battery.

This apparatus, which is suitable for measuring feeble electrifications, will enable us to proceed to some fundamental experiments.

80. Experiments.—Let us consider a metallic cylinder, *A*, Fig. 48, open at the top and supported on an insulating

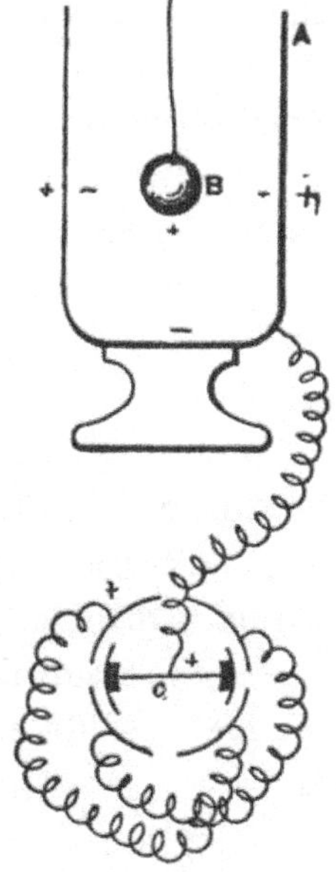

FIG. 48.

base; let us also connect the cylinder with the framework, *c*, in the electrometer by a thin wire, both pairs of quadrants being electrified in the manner indicated in the preceding paragraph.

If the cylinder be in the neutral state, the framework will not change its position.

I. Introduce into the cylinder a conductor, B, hung from a silk thread and charged with positive electricity. We know, by the phenomenon of influence, that the internal wall of the cylinder will be charged negatively, and that the electricity of the same sign as that of B will be repelled towards the external wall and to the framework, c. This latter will exhibit a displacement which will increase as the conductor B descends, until this last has reached a certain depth in the cylinder. From this point onwards the displacement of the framework c remains invariable, whatever be the position of B; even if B be put in contact with the cylinder, the displacement of the electrometer remains the same. Upon withdrawing the body after contact it will be found, by means of a pith-ball electroscope or a second quadrant electrometer, that B has returned to the neutral state. Hence we conclude that the positive charge which it has parted with to the cylinder has been neutralized by the negative charge developed by influence, the cylinder preserving its induced positive charge.

If the charge on B were negative, the electrometer would have shown a displacement in the opposite direction.

II. Place inside the cylinder a metallic sphere charged with electricity and giving a displacement α. Touch this sphere with an equal one in the neutral state. Then the two spheres, if placed successively inside the cylinder, will give deviations equal to $\alpha/2$.

III. Suspend inside the cylinder two insulated bodies in the neutral state and rub them one against the other. The electrometer will show no deviation, but will remain in its first position.

IV. If we introduce separately into the cylinder various bodies, B, D, E, charged with electricity, we get deviations some of which are positive, others negative, according to the signs of the charges.

On introducing simultaneously B, D, and E the deviation obtained is the algebraic sum of the preceding deviations and remains invariable, even if the bodies are put in contact or rubbed against each other.

From these experiments we conclude that electric charges are quantities susceptible of measurement. We might take a given charge as unity and consider as double and triple those charges which produce a double or triple deviation in the electrometer connected with the cylinder.

The two last experiments show that the total charge of a system of electrified bodies is invariable, and that friction produces on bodies equal and opposite electricities, capable of neutralizing each other.

If, for example, we electrify a resin rod by aid of a piece of cloth, a quantity of electricity equal and opposite to the charge on the resin is produced on the cloth and on the neighboring conductors connected with it, such as the body of the experimenter, the table used for the experiment, and the walls of the room.

81. In the case of a conductor in equilibrium the electricity is distributed on its surface. — This important property can be demonstrated in various ways.

I. For example, let A, Fig. 49, be a hollow metallic sphere. Touch the exterior surface of the sphere with a proof-plane formed of a disc of copper foil on an insulating handle; the disc will take an electric charge, as can be discovered by means of an electroscope or electrometer. If, on the other hand, the point touched be on the interior surface of the sphere, the disc will show no electrification.

The method of the proof-plane, though by no means exact, can be employed to compare the electric charges on the different parts of a body. If we touch successively various points on the surface of an electrified sphere, we find that

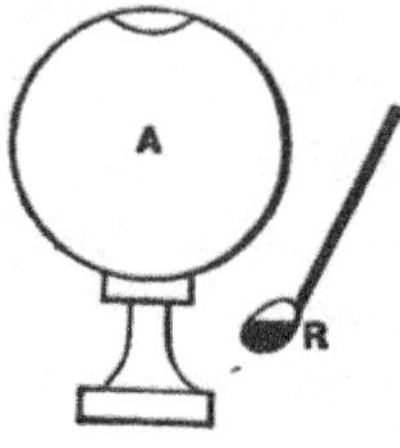

FIG. 49.

the charge carried away each time is constant, that is, that the sphere is uniformly charged. For a body of ovoid form we find that the charge increases inversely as the radius of curvature of the surface.

II. The fact of the external distribution of the charge on a conductor has been established beyond doubt by Faraday, who had a chamber four metres square constructed, supported on insulators and covered with metallic sheets. He entered this chamber while it was electrified and was unable to discover the least trace of electricity on its internal walls, though using the most delicate electrometers.

82. Law of Electric Actions.—The preceding facts, on being summed up, show that the action between electrified bodies must be placed amongst the central forces, and that we can define by *quantity of electricity*, or electric mass, or charge of a body, a quantity proportional to the force which it exercises upon neighboring electrified bodies.

The superficial density is the charge per unit surface.

In virtue of this definition the elementary law of the force between two electric masses q and q', at a distance l, will be of the form

$$f = kqq'\phi(l).$$

The function $\phi(l)$ is determined by the fact that the electric action is zero in the interior of an electrified conductor in equilibrium, § 81.

The reciprocal of the theorem demonstrated in § 26 shows that, in order that a homogeneous spherical shell may have no action upon an internal point, the force must be in inverse ratio to the square of the distance.

The law of electric attraction and repulsion is therefore

$$f = k\frac{qq'}{l^2},$$

in which we consider the force f as repulsive or attractive according as the masses q, q' are of the same or contrary sign.

This law, experimentally demonstrated by Coulomb, enables us to apply to electric attraction the general properties of the Newtonian central forces.

83. Definitions. Electric Field. Electric Potential.—

By *electric field* we will designate a space or region where electric forces take origin. The *intensity* in a point of an electric field is the resultant of the forces exercised there upon a positive mass taken as unity. The *direction* of this resultant is called the direction of the field.

Electric forces are defined by aid of a potential, § 12, called *electric potential*, which has as its expression

$$U = k\Sigma\frac{q}{l}.$$

The intensity of the field in a direction s is expressed as a function of the potential by

$$\mathcal{K}_s = -\frac{du}{dl}.$$

84. Potential of a Conductor in Equilibrium.—

Since the force is zero at the interior of a conductor carrying an electric charge in equilibrium, § 81, the electric potential has the same value at all internal points; the surface of the conductor is equipotential, and the lines of force are normal to its surface.

This constant value is called the potential of the conductor.

The idea of potential may be distinguished from the idea of charge by the following experiment: The method of the proof-plane demonstrates that the charge is variable at different points of a conductor of irregular shape. Nevertheless, if we successively connect these points by a wire with the electrometer, the deviation remains constant, for the force which urges the electricity to pass from the conductor to the electrometer depends only on the potential, which is invariable.

In the case of a metallic sphere of radius R electrified to a density σ the potential at the centre is

$$U = k\frac{Q}{R} = k\frac{4\pi R^2 \sigma}{R} = 4\pi k R\sigma, \qquad \text{§ 26.}$$

This quantity is the potential of the sphere.

85. Potential of the Earth.—Electric forces do not depend at all on the absolute values of the potential, but only on its variation. Thus the force which impels the electricity of a conductor to flow to an adjoining conductor, such as the earth, depends on the difference between the potentials of the bodies under consideration; positive electricity tends to be displaced in the direction of decreasing potentials. The ground and the walls of an experimenting-room being conductors, there is no objection to considering their potential as zero and then taking as positive those potentials above that of the earth, the potentials being negative in the opposite case.

86. Coulomb's Theorem.—*The intensity of the field at a point infinitely near a conductor in equilibrium is equal to $4\pi k$ multiplied by the superficial density at that point.*

Let us consider an element ds of the surface charged with a quantity of electricity equal to σds, σ being the density.   Let us suppose a tube of force, passing through the contour-line of ds, limited externally by an infinitely near equipotential surface U, and closed inside the conductor by any surface S.   Let us then apply Gauss's theorem, § 20, to the volume thus enclosed.   The flux of force is zero across the surface S, since there is no electric force inside a conductor in equilibrium; as there is no component across the lateral walls of the tube, the outflowing flux from the given volume is limited to the flux across the equipotential surface U_1.

Let $\mathcal{H}_e$ be the intensity of the field normal to this surface; the flux $\mathcal{H}_e ds$ is equal to $4\pi k$ multiplied by the quantity of electricity σds.

Consequently

$$\mathcal{H}_e = 4\pi k\sigma.$$

The intensity of the field having the same sign as the density, it is directed outwards for a positive density, inwards for a negative destiny.

We also have

$$\mathcal{H}_e = -\frac{dU}{dn},$$

U being the potential of the conductor, n a direction normal to the surface.

We see that the property demonstrated in the case of a sphere, § 27, is general for all electrified conductors in equilibrium.

87. Electrostatic Pressure.—The method of reasoning applied to the end of § 30 shows that the force exercised by the whole charge of a conductor in equilibrium upon the electricity covering unit surface is equal to $2\pi k$ multiplied by the square of the density at the given point:

$$P = 2\pi k\sigma^2.$$

This force, called electrostatic pressure, is always positive whatever be the sign of σ. Hence it results that the electricity has a tendency to escape from the conductor and enter the surrounding medium. If the conductor is movable, it may be moved in the direction of the maximum pressure. Thus, when two conductors are electrified with opposite signs, their electric density is, by the phenomenon of induction, greater on the neighboring faces than on the opposite ones; the electrostatic pressure therefore tends to urge the conductors towards each other.

When the bodies are charged with electricity of the same sign, the maximum densities are on the outer faces and the resultant pressures are in such a direction as to move the conductors apart. So, too, if a soap-bubble be electrified, it expands until the increased surface-tension of the liquid layer balances the electrostatic pressure.

88. Corresponding Elements.—Suppose a tube of force, traversing an electric field, be limited by two electrified conductors on which it cuts surfaces s, s' called *corresponding elements.* Let q and q' be the charges of these surfaces. Then, applying Gauss's theorem, § 20, to the volume limited by the tube and by any two surfaces inside the conductors, it is easy to see that the outflowing flux is zero, since the electric force has no component inside the conductors nor across the lateral walls of the tube.

Consequently

$$4\pi k(q + q') = 0;$$

whence

$$q = -q'.$$

We conclude from this that *corresponding elements carry equal and opposite quantities of electricity.* The lines of force uniting them take their origin from the points charged with positive electricity and end in points covered with negative electricity.

The lines of force which leave an electrified body necessarily find, therefore, an opposite electric charge in their vicinage. This charge is spread over the neighboring conductors, over the table which supports the body, over the walls of the experimenting-room. This is the true method of looking at the phenomenon of induction, § 78, the lines of force which emanate from the inducing body impinging on the influenced body.

89. Power of Points.—If the inducing body is furnished with a point, the lines of force emerging from this latter form a conical tube. The apex of the cone carries a charge equal to that of the base, so that the density acquires an excessive value at the point, and the electrostatic pressure urges the contrary electricities to recombine across the medium which separates them.

90. Electric Screen.—A hollow conductor surrounded by electrified bodies will assume on its surface a distribution of induced electricity; its potential will acquire a constant value equal to the potential of all the internal points, § 84. Consequently no electric force will exist inside the conductor, and any objects that it may contain will be entirely cut off from the action of the exterior charges. The hollow conductor thus serves as a *screen* to the bodies it envelops. Faraday's experiment, § 81, is a confirmation of this property.

Maxwell has shown that the screen need not necessarily be continuous; a metallic gauze constitutes an efficacious screen.

Similar means of protection are used to cut off electrometers from external influences.

91. Lightning-rods.—The two properties just considered are made use of in the preservation of buildings from lightning. The best lightning-rod would be a metallic covering completely enveloping the whole building. As this is hardly practicable, we may adopt, as was done by Melsens (in Brussels), the plan of covering the building with a network of conductors connected with all the metallic portions of the building.

This network is connected to earth by means of the gas- and water-pipes and by metal plates sunk in ponds or moist earth. On the top points of the building are fixed pointed rods or clusters of spikes of metal connected to the general network.

CONDENSERS—DIELECTRICS.

92. Capacity of Conductors.—Suppose an insulated conductor A to be in the neighborhood of other conductors B, C, D, maintained at a constant potential by some arrangement, for example, by connection with the earth. If we give A an electric charge Q, the other conductors assume induced charges which are distributed according to a law depending on the form of these bodies, on their relative positions, and, as we shall see, on the medium which separates them. The sum of these charges is equal and contrary to Q. The potential of A will be expressed by

$$U = k \Sigma \frac{q}{l}.$$

If we increase the charge on A, the induced charges increase in the same proportion; the new induced layers are superposed on the first, and if the relative distances remain the same the potential of A grows in proportion to its charge.

Consequently the ratio between the charge of A and its potential is a constant which only depends on the shape and relative position of the conductors composing the system, and on the medium which separates them.

This constant,

$$C = \frac{Q}{U},$$

is called the *capacity of the conductor A.*

If $U = 1$, $C = Q$; the capacity of a body is therefore measured by the charge which raises its potential one unit above the potential of the surrounding conductors. Nevertheless the capacity is not a charge, any more than the content of a vessel for holding a liquid represents a quantity of the liquid.

93. Condensers.—Spherical Condenser.—As an application of the preceding, let us consider a metallic sphere enveloped by a concentric hollow sphere, this latter being connected to earth.

Such an arrangement constitutes a *condenser*, the spheres being the *armatures* (or *plates*) and the insulator which separates them the *dielectric*.

When a charge $+q$ is given to the interior sphere, the internal wall of the other sphere takes, by induction, a charge equal and opposite to the first, § 88, and the electricity of the same name as that of the interior sphere escapes to the earth.

If we denote by l and l' the radii of the two electric layers, the potential at the centre of the two spheres is

$$U = k\left(\frac{q}{l} - \frac{q}{l'}\right) = q\frac{k(l' - l)}{ll'}.$$

According to the definition given above, § 92, the capacity of the interior sphere, called also the *capacity of the condenser*, is

$$C = \frac{q}{U} = \frac{ll'}{k(l' - l)}.$$

We see that the capacity of a condenser increases inversely to the distance between its plates.

If the radius l' were to increase indefinitely, we should have at the limit an insulated sphere whose capacity would be

$$C = \frac{l}{k}.$$

94. Plate Condenser.—Let us take the case of a condenser formed of two parallel discs, with rounded edges, A, B, A being electrified, while B is connected with the earth. The adjoining faces of A and B take equal and opposite charges of electricity, and the lines of force run perpendicularly from one face to the other, except at the edges, where the lines curve in such a way as to remain perpendicular to the surfaces. We can therefore admit that in the central region the electric field is uniform.

The intensity of the field is expressed by

$$\mathcal{K}_e = -\frac{dU}{dn},$$

n being the direction of a line of force. . As the intensity is constant in this direction, we have, representing by U the potential of A and by l the distance between the plates,

$$\mathcal{K}_e \int_0^l dn = \int_U^0 - dU;$$

whence

$$\mathfrak{K}_e l = U \quad \text{and} \quad \mathfrak{K}_e = \frac{U}{l}.$$

But the field is also expressed by $\mathfrak{K}_e = 4\pi k\sigma$, § 86, σ being the surface-density.

Consequently

$$\frac{U}{l} = 4\pi k\sigma \,;$$

whence

$$\sigma = \frac{U}{4\pi kl}.$$

The charge on a surface s, taken in the middle region, will be

$$q = \frac{Us}{4\pi kl}.$$

The capacity of the condenser referred to this surface will be expressed by

$$\frac{q}{U} = \frac{s}{4\pi kl}.$$

We have neglected the charge taken by the plate A on its posterior face; but if l is small this quantity is negligible compared with that on the anterior surface.

A condenser of great capacity is obtained by superposing leaves of tinfoil, separated by sheets of paraffined paper or of mica and connected in two series, the first comprising the even-number sheets, the second comprising the odd-number ones. The sheets of such a condenser cannot be charged at very different potentials for fear of piercing the dielectric by sparking.

95. Guard-ring Condenser.—Lord Kelvin has devised a uniformly charged plane plate by cutting a narrow circular groove through the plate, A, Fig. 50, and electrically connecting the interior disc with the exterior ring, called guard-ring, by a slender wire. It can be admitted that the

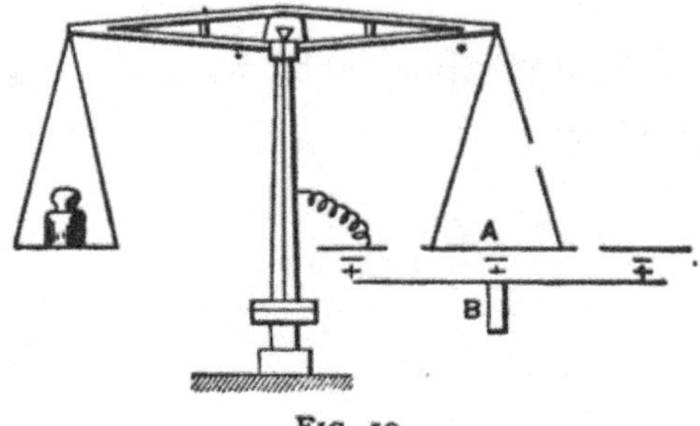

FIG. 50.

circular cut has not sensibly modified the uniformity of the field, so that the capacity of the disc A, of surface s, is rigorously

$$\frac{s}{4\pi kl}.$$

96. Absolute Electrometer.—The pressure which is exercised by unit surface on the disc A is, § 87,

$$2\pi k\sigma^2;$$

the pressure on a surface s will be

$$p = 2\pi k\sigma^2 s.$$

Now

$$\sigma = \frac{U}{4\pi kl};$$

whence

$$p = \frac{U^2 s}{8\pi kl^2}.$$

If we support the disc A by the beam of a balance, Fig. 50, we will be able to balance the attraction of the two discs by a known weight. Expressing this weight in dynes, we can deduce from the preceding equation the difference of potential of the two plates:

$$U = \sqrt{\frac{8\pi k p l^2}{s}}.$$

The above is the principle of Lord Kelvin's absolute electrometer.

97. Cylindrical Condenser.—Suppose a condenser formed of two indefinite concentric cylinders. The interior cylinder, of radius r_1, having a charge q, the outer cylinder, of radius r_2, connected to the earth, will take a charge $-q$. Such a condenser is realized in practice by a conducting wire covered with an uniform insulating layer and submerged in water, this latter serving as the outer plate.

The dielectric can be divided by equipotential surfaces which are concentric with the conductor by reason of symmetry. The tubes of force are limited by planes passing through the axis of the cylinders.

Let us take the case of a tube of opening α, limited by two planes normal to the axis and one centimetre apart.

The flux of force is constant in the tube, § 21.

Denoting by s a section of the tube by an equipotential surface of radius r, we have

$$\mathcal{K}_r s = \mathcal{K}_r \times \alpha r = \text{const.},$$

whence, denoting by $\mathcal{K}_1$ the intensity of the field infinitely near the interior plate, we get

$$\mathcal{K}_r r = \mathcal{K}_1 r_1 = 4\pi k \sigma r_1, \quad § 86.$$

The capacity per unit of length is

$$C = \frac{q}{U} = \frac{2\pi r_1 \sigma}{U}.$$

But

$$\mathfrak{K}_e = -\frac{dU}{dr} ;$$

whence

$$\int_U^0 - dU = \int_{r_1}^{r_2} \mathfrak{K}_e dr = \int_{r_1}^{r_2} \frac{4\pi k\sigma r_1 dr}{r} ,$$

which gives

$$U = 4\pi k\sigma r_1 \, \log_\epsilon \frac{r_2}{r_1} ;$$

and consequently

$$C = \frac{q}{U} = \frac{1}{2k \log_\epsilon \dfrac{r_2}{r_1}} .$$

This expression shows that, when we connect condensers with a source of electricity, we must avoid using insulated wires in contact with conducting surfaces, for a condensation is then produced between the wires and the surfaces which may make our results incorrect. It is best to employ slender wires, at a considerable distance from conducting surfaces, so as to obtain a negligible capacity.

98. In the preceding calculations we have supposed one of the condenser-plates to be connected to earth, so that it preserves a constant potential, which in this case is zero. If the plates were given potentials U_1 and U_2, the capacity would be expressed by

$$C = \frac{q}{U_1 - U_2}.$$

99. **Leyden Jar.**—If in the formula for a spherical condenser we suppose the radii r and r' but slightly different, we have approximately

$$C = \frac{r^2}{kb},$$

b being the distance between the spheres; whence

$$C = \frac{4\pi r^2}{4\pi kb} = \frac{s}{4\pi kb}.$$

This formula, identical with that for a plane condenser, is applicable to a condenser of any form whatever, provided that the plates be sufficiently near together and their curvature sufficiently slight for us to consider the field inside the dielectric as uniform.

When great exactness is unnecessary, this is allowed to be the case in the *Leyden jar*. Contrary to the tinfoil condenser, described in § 94, this condenser can sustain considerable differences of potential. It consists of a bottle or jar of glass varnished with shellac, and whose inner and outer surfaces are covered with tinfoil up to a certain distance from the mouth of the jar. A metal rod, passing through the neck and terminating in a knob, gives a means of connecting the inner coating with a source of electricity.

In order to obtain great capacity a number of jars can be connected together, the inside coatings being joined to each other, and the outside coatings also connected by themselves.

100. Energy of Electrified Conductors.—Let $q, q', q'' \ldots$ be a system of electric masses at potentials $U, U', U'' \ldots$

In virtue of the properties demonstrated for Newtonian forces, § 25, the potential energy of the system is

$$W = \tfrac{1}{2}\Sigma qU.$$

It will be noticed that the conductors connected to earth, whose potential is zero, do not furnish any element to the above sum, even when they carry electric charges.

In the case of a condenser one plate of which is connected to earth, q and U being respectively the charge and potential of the other plate,

$$W = \tfrac{1}{2}qU.$$

If we denote by C the capacity of the condenser,

$$q = CU\,;$$

whence

$$W = \tfrac{1}{2}qU = \tfrac{1}{2}CU^2 = \frac{1}{2}\frac{q^2}{C}.$$

101. Theory of the Quadrant Electrometer.—The cylindrical framework of the quadrant electrometer (which for the sake of simplicity we shall call the *needle* of the electrometer), Fig. 51, constitutes a condenser with each of the pairs of quadrants, these latter being connected two and two.

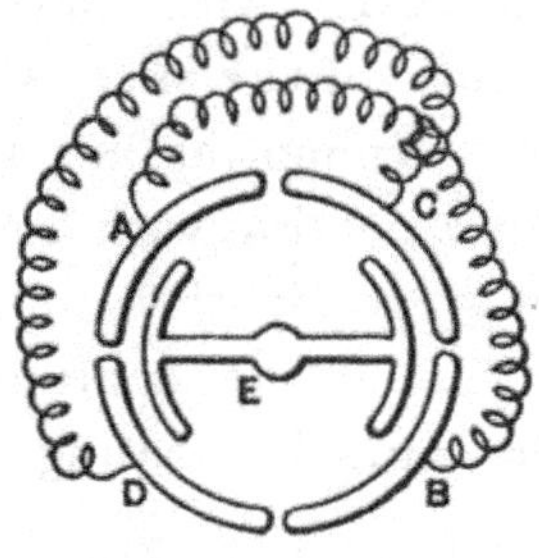

FIG. 51.

Call U the potential of the needle,

U_1 that of the quadrants $A, B,$

U_2 that of the quadrants $C, D.$

The charge of the condenser $E - A, B$ is proportional to the difference of the potentials $U - U_1$, and the horizontal component of the force tending to displace the needle towards the given quadrants can be represented by $a(U - U_1)^2$, a being a constant coefficient, § 87.

The needle is drawn in the opposite direction by a force equal to $a(U - U_2)^2$, the constant a being the same, in consequence of the slight displacement of the needle and the identity of shape of the quadrants.

The resultant of these two actions is

$$a(U_2 - U_1)(2U - U_2 - U_1).$$

Its moment is counterbalanced by the torsion-couple of the suspending wire. Coulomb has shown that, within certain limits, this couple is proportional to the angle of torsion.

We may therefore write

$$\beta = b(U_2 - U_1)\left(U - \frac{U_2 + U_1}{2}\right).$$

This formula shows that the deviation is zero when the quadrants are at the same potential.

102. I. When the potential of the needle is very large relatively to that of the quadrants, the formula reduces to

$$\beta = b(U_2 - U_1)U.$$

II. If we connect the needle to one of the pairs of quadrants, CD, for example, we have

$$U = U_2;$$

whence

$$\beta = b(U - U_1)\left(V - \frac{U + U_1}{2}\right) = \frac{b}{2}(U - U_1)^2.$$

The deviation is then proportional to the square of the difference of potential of the needle and A, B.

III. In the case where the quadrants are maintained at equal and opposite potentials, $U_1 = - U_2$, which can be done, as we shall see, by connecting them to the poles of a battery, we have

$$\beta = 2bUU_2.$$

The deviation is proportional to the potential of the needle.

M. Gouy* has shown that the preceding formula does not hold when the potentials of the quadrants are considerable relatively to that of the needle. In such case there exists a directive electric couple acting together with the torsion-couple, and whose action must be taken into account.

It must be further remarked that, if the needle is not of the same metal as the quadrants, there arises a difference of potential by contact, § 107, which slightly alters the results, and which is eliminated by making two successive experiments in which this extraneous difference of potential produces equal and opposite effects. The mean of the deviations thus obtained is the result desired.

103. Specific Inductive Capacity of Dielectrics.— We have so far paid no attention to the dielectric separating the condenser-plates. For the purpose of studying this element let us consider two identical plane condensers, the dielectric in the one being air, and in the other paraffine.

The similar plates being connected to earth, electrify the free plate of the first condenser to the potential U, and connect it with the needle of an electrometer whose quadrants are supposed to be at equal and contrary potentials. Let α be the deviation of the needle. Then connect the free plates of the two condensers by a wire. As the initial charge q spreads over both condensers the potential diminishes and becomes U'; the deviation of the electrometer falls to α'. Calling c and c' the capacities of the two condensers, we get the condition

$$q = cU = (c + c')U';$$

whence

$$\frac{U}{U'} = \frac{c + c'}{c}.$$

---

* *Journal de Physique*, 1888.

$$\frac{}{c} = \frac{}{a'} - 1.$$

The ratio $\dfrac{c'}{c}$ is called the *specific inductive capacity* of the paraffine in comparison with air. This ratio is about 2.3 in the present case.

We might compare the capacities of dielectrics to that of a vacuum by placing one of the condensers in a vacuum-chamber. We give here some of the values thus obtained, according to Boltzman:

Sulphur...................... 3.84
Glass........................ 5.83 to 6.34
Paraffine 2.32
Ebonite...................... 2.21 to 2.76
Essence of turpentine........ 2.21
Air.......................... 1.00059
Carbonic anhydride........... 1.000946

The comparison of these values with the indices of refraction i and i' of the same substances for light shows that the former are sensibly proportional to the squares of the latter.

As the index of refraction of a medium is inversely proportional to the velocity of light in it, it follows that the specific inductive capacities of bodies are inversely proportional to the square of the velocity of light in those bodies:

$$\frac{c'}{c} = \frac{i'^2}{i^2} = \frac{v}{v'^2}.$$

104. Nature of the Coefficient k in Coulomb's Law.— Given an air-condenser charged with a quantity of electricity q at a potential

$$U = k\Sigma\frac{q}{l},$$

the capacity is

$$c = \frac{q}{U}.$$

Let us substitute for air some other dielectric; keeping the charge constant, we will get a new capacity,

$$c' = \frac{q}{U'},$$

the potential necessarily varying and becoming

$$U' = k' \Sigma \frac{q}{l}.$$

Hence we deduce

$$\frac{c}{c'} = \frac{U'}{U} = \frac{k'}{k} = \frac{v'^2}{v^2}.$$

Now the ratio of the capacities $\frac{c}{c'}$ is the same as that of the specific inductive capacities.

Consequently the coefficient in Coulomb's law is inversely proportional to the inductive capacity of the dielectric separating the electrified bodies, and, if the deductions of the last paragraph are true, this coefficient is also proportional to the square of the velocity of light in the dielectric.

It is because the coefficient k is not an arbitrary factor, but an actual physical quantity, that we have retained it in our formulæ. The importance of this coefficient will be seen in the comparison of electrical units.

105. Rôle of the Dielectrics. Displacement.—What we have just seen shows the importance of the rôle of the dielectric in electrostatic actions. We present here an experiment which leads to the consideration of the phenomenon of condensation (or *accumulation*) in a new light.

A condenser formed of three detachable parts, a glass jar or plate and two metallic plates, is charged in the ordinary way. The three parts are then separated and the metal plates connected to earth ; then the condenser is put together again, and it can be discharged as if it had remained undisturbed.

This experiment shows that the charge is, at least in part, on the dielectric.

Faraday has deduced an ingenious hypothesis from this fact. According to him, dielectrics present a polarization which recalls that which we have met with in the phenomena of magnetization. This polarization takes place along the lines and tubes of force which join the plates of condensers and electrified bodies in general.

Take the case of an elementary tube between two conductors A and B joining corresponding elements ds_1, ds_2 charged with densities σ_1, σ_2, § 88.

This tube can be subdivided by equipotential sections into elements of volume presenting equal and contrary charges on their opposing ends, thus balancing each other. At the extremities of the tube exist quantities of free electricity,

$$+ q = \sigma_1 ds_1 ,$$
$$- q = \sigma_2 ds_2 ,$$

which charge the conductors, these latter having no other function than to limit the dielectric.

In this hypothesis the product $\sigma ds = \sigma_1 ds_1$ is constant along the tube ; but at the surface of the conductors, § 86, we have

$$\sigma_1 = \frac{\mathcal{H}_e}{4\pi k} ;$$

and as the product $\mathcal{H}_e ds = \mathcal{H}_1 ds_1$ is constant in the tube, § 21, we conclude that in any point of the field the quantity of electricity displaced per unit equipotential surface is

equal to the intensity of the field at this point divided by
the factor $4\pi k$:

$$\sigma = \frac{\mathfrak{K}_e}{4\pi k}.$$

Maxwell calls this expression the *displacement of electricity*
in the dielectric.

The analogy which exists between electric induction thus
interpreted and magnetic induction will be recognized:
electric displacement corresponding to intensity of magnetization, and the specific inductive capacity to the coefficient
of magnetization (*magnetic susceptibility*).

This point of view gives us an explanation of the energy
of electrified conductors; we conceive, namely, that there
results from the polarization of the dielectric molecules a
state of tension along the lines of force, in virtue of which
these lines tend to shorten themselves by causing the
approach of the conductors which limit the dielectric.
This tension, which is comparable with that of an elastic
body, assimilates the energy of the dielectric to that of a
spring.

It can be readily verified that, if the electric field is
uniform between the plates of a condenser, such as a guard-
ring condenser, the total energy is expressed by the volume
of the dielectric multiplied by $\dfrac{\mathfrak{K}_e^2}{8\pi k}$, $\mathfrak{K}_e$ being the intensity of
the field.

In fact, keeping the previous notation,

$$W = \tfrac{1}{2}qU = \tfrac{1}{2}\sigma s \times \mathfrak{K}_e d = \tfrac{1}{2}\mathfrak{K}_e \sigma \times \text{volume}.$$

Now

$$\mathfrak{K}_e = 4\pi k\sigma;$$

whence

$$W = \text{volume} \times \frac{\mathfrak{K}_e^2}{8\pi k}.$$

In the case of a uniform field the energy of the dielectric per unit volume is therefore expressed by

$$\frac{\mathcal{H}_e^2}{8\pi k}.$$

Maxwell has shown that this holds in the case of any field whatever, and that the law of the attractions and repulsions of electrified bodies can be established by admitting, along the lines of force, a tension of $\dfrac{\mathcal{H}_e^2}{8\pi k}$ per unit of surface normal to the field.*

106. Residual Charge of a Condenser.—The polarized condition of a solid dielectric does not always cease after the discharge of the plates.

Suppose we charge a Leyden jar, the inner coating of which is connected with the needle of a quadrant-electrometer, while the outer coating is connected with the earth. At the moment of discharging the jar the needle returns to zero, but gradually reassumes a deviation but slightly different from the first. The discharge may be repeated a certain number of times before the needle finally rests at zero.

This phenomenon, which presents some analogy with the remanent magnetism of a bar of steel, is particularly observable in the case of dielectrics which are not pure, such as glass, gutta-percha, caoutchouc; sulphur and paraffine exhibit this phenomenon to a much smaller extent.

The electrification of the plates after a first discharge is called the *residual charge*. This varies with the duration of the electrification of the condenser, while the *effective charge* that can be obtained at the moment of the first contact between the plates is independent of this duration. For this reason it has been agreed to call the ratio of the *effective*

---

* Maxwell, *Electricity and Magnetism*, Vol. I, Chap. V, etc.

charge to the difference of potential of the plates the capacity of the condenser.

From the preceding considerations it follows that, for the same difference of potential of the plates, the charge of a condenser is greater with decreasing than increasing differences of potential. The electrization of a dielectric placed in an electrical field of varying intensity presents, it will be seen, a true parallel with the magnetization of iron introduced into a field of varying intensity, which has resulted in the name of *dielectric hysteresis* being given to the phenomenon here considered. In both cases the cyclical variations of the intensity of the field causes heating.

107. Electromotive Force of Contact. Distinction between Electromotive Force and Difference of Potential.—Volta discovered that two metals, e.g., zinc and copper, assume a definite difference of potential on contact.

Following Lord Kelvin's experiment, let us make a disk, one half being of copper and the other of zinc, and hang a slender sheet of metal over the line of contact. When this sheet is positively electrified, it is displaced towards the copper; electrified negatively, it is attracted towards the zinc.

We might also make the electrometer of § 79 with two quadrants, A, B, of zinc, and the other two, C, D, of copper On connecting the zinc and copper quadrants by copper wire we will get a deviation of the needle towards the zinc or towards the copper, according as its charge is negative or positive.

This experiment shows that upon contact the zinc is positively charged, the copper negatively. In virtue of these charges the two metals take opposite potentials, whose difference is called the *electromotive force of contact.*

Similar electromotive forces probably arise on the contact of all dissimilar bodies. Helmholtz supposes them to be due to the different specific attractions of the bodies for electricity. Those whose attraction is greater take electricity in excess, that is, are positively charged. At the surface of separation are produced two opposite layers of electricity, occasioning the observed difference of potential.

This hypothesis accounts for the fact, shown by experiment, that the same body is capable of taking opposite charges, according to the nature of the bodies placed in contact with it. Copper, for example, is negative with regard to zinc and positive with regard to silver.

In a general way, the name electromotive force is reserved for every *cause* of displacement of electricity; a difference of potential is the *effect* of an electromotive force. This distinction is analogous to that between the pressure exercised on a liquid and the difference of level which results therefrom in a manometer.

The electromotive force of contact gives an explanation of the development of electricity by friction, which, by this hypothesis, has no other effect than to render the contact of the rubbed bodies more thorough.

The high potential acquired by bodies when rubbed is explained by the separation of the opposite charges. Suppose, for example, two discs, one of zinc, the other of copper, in contact. The potentials assumed by these two metals depend on their charges and relative distances; e.g., for a point on the zinc disc the potential, which is constant in the whole homogeneous mass, is given by the sum of the ratios of the positive charges of the disc to their distances from this point, less the sum of the ratios of the negative charges, distributed over the adjoining disc, to their corresponding distances.

If the two discs are moved apart, the subtractive term diminishes very rapidly, so that the positive potential of the zinc becomes considerable, as does also the negative potential of the copper.

The increase in potential energy,

$$W = \tfrac{1}{2}\Sigma qU,$$

which results is equal to the work expended in separating the opposite charges.

ELECTRIC DISCHARGES AND CURRENTS.

So far we have been studying the properties of conductors charged with layers of electricity retained by dielectrics and supposed to be in equilibrium.   Let us now take up the case of rupture of equilibrium and the manifestations resulting from that rupture.

108. Convective Discharge.—The outside coating of a charged Leyden jar, Fig. 52, is connected with a metallic knob b', in proximity to the knob b of the inner coating.   A pith-ball s, being hung by a silk thread between the two knobs, is electrified by influence and attracted by the nearest

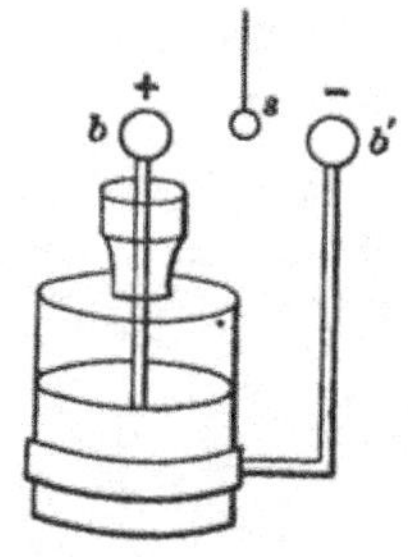

FIG. 52.

knob, on coming into contact with which it takes a charge of the same sign.   It then swings to the opposite knob and continues to oscillate from one knob to the other, carrying small electric charges, until the opposite electricities of the coatings are neutralized.

This phenomenon may be interpreted as a transporting of positive electricity from b to b', and of negative electricity in the opposite direction. On the hypothesis of a single fluid, the whole transportation is from b to b'.

The potential energy of the jar, $\frac{1}{2}\Sigma qU$, is gradually reduced to zero; it is used up in the blows of the swinging ball against the knobs b and b' and the friction of the ball against the air; a transformation of electric energy into heat therefore takes place.

This discharge of the jar is called *convective*, because the displacement of electricity is connected with the movement of the body which conveys it.

109. Conductive Discharge. Electric Current.—The contrary electricities which charge the coatings are made to recombine directly, if they are joined together by a conductive wire. The electrification is then observed to disappear instantaneously without displacement of the conductor, but we can show that also in this case there is a transformation of electric energy into heat in the conductor.

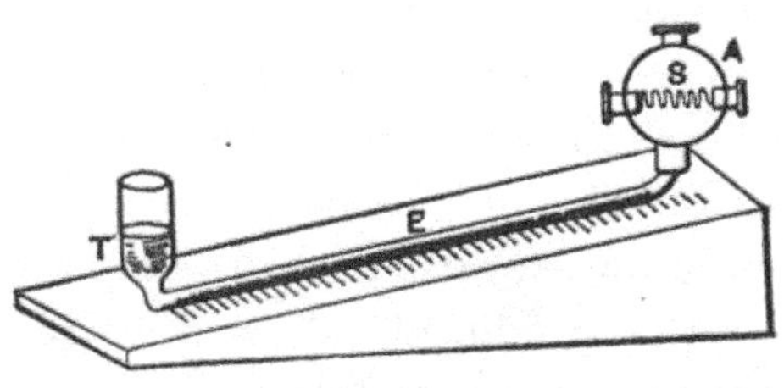

FIG. 53.

For this purpose Riess' thermometer is used, which consists of a glass bulb A traversed by a platinum spiral s. The bulb communicates with an inclined glass tube E, terminating in a vertical tube T which contains a colored liquid. When a discharge is passed through the platinum spiral, the heat developed in it is communicated to the air in the bulb

and causes a sudden displacement of the liquid in the inclined tube; the expansion of the air enables us to calculate
the heat given off.

If we neglect the loss by radiation, then, denoting by c
the specific heat of platinum, m the mass of the wire, c', m'
the same terms for the air in the bulb, d the alteration of
level of the liquid, θ and θ' the initial and final temperatures,
the heat developed is

$$W = (mc + m'c')(\theta' - \theta) = ad,$$

a being a numerical coefficient.

The mechanism of the conductive discharge is explained,
like that of the convective, by supposing that the molecules
of the conductor play the same part as the pith-ball; ex-
changes of electricity take place from molecule to molecule
by the displacement of these latter. The increase in molecular activity corresponds to the observed development
of heat.

It must not, however, be forgotten that before the introduction of the conductor between the coating the seat of
the energy was in the dielectric, § 105. The transfer of
energy from the dielectric to the conductor implies that at
the moment of establishing the metallic connection the lines
of electric force crowd together into the conductor, in which
a polarization is thus produced, followed by transfers of electricity due to the molecular agitation.

The mechanism of this transfer still, however, remains
to be explained. We shall see, when studying variable currents, that the development of heat is first produced in the
outer parts of the conductor, and that when the phenomenon
is sufficiently rapid it cannot reach the middle of the
conductor.

Another way of looking at the matter, less exact, perhaps,

but more convenient, suggests the ideas of hydrodynamic phenomena, consists in supposing a simple transportation or *current of electric fluid* in the conductor joining the coatings taking place from the positive knob to the negative one. The heat produced in the conductor is then explained in the same way as that caused by a ponderable fluid rubbing against the sides of a pipe; the movement of the electric fluid is communicated to the molecules of the conductor which it thus heats.

The quantity of electricity which passes per second at any given time across the section of the conductor is called the *strength of current* (or simply *the current*). It has been agreed to assume that the current is directed from the positive to the negative plate.

110. Disruptive Discharge. Electric Spark and Brush. Their Effects.—If we adopt Faraday's and Maxwell's view, we may suppose that, upon gradually increasing the difference of potential of the plates of a condenser, there comes a time when the tension of the dielectric is no longer balanced by its tenacity, so that a rupture is produced; this is, in fact, shown by experiment. The dielectric of a condenser is pierced when the tension exceeds a certain limit, and there is then a sudden return to the neutral state, while at the same time a spark is observed to pass between the plates; this discharge is called *disruptive*.

In gases the sparking takes place at a much lower tension than in solid dielectrics. The length of the spark for a given tension naturally depends on the shape of the conductors used, since the electrostatic pressure varies as the square of the density on the conductors, § 87. It varies also with the pressure of the gas; compression diminishes the explosive distance, while rarefaction increases the length of the spark up to a certain limit, beyond which the decrease is rapid. In a vacuum the disruptive discharge completely ceases.

Matter is therefore necessary for the transfer of electricity, which seems to show that it is the material molecules which serve as its vehicle, as in the case of convective discharge.

In a gas under constant pressure and with pointed electrodes the length of the spark grows much more rapidly than the potential difference of the conductors ; a fact which leads us to think that the large sparks due to storm-clouds are not caused by potentials out of proportion to those of our electrostatic machines.

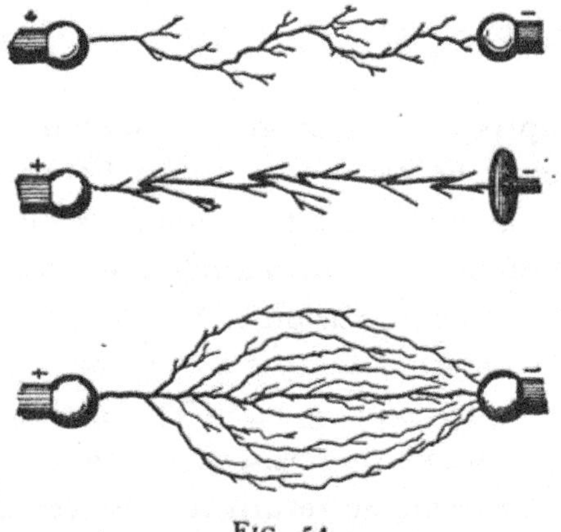

FIG. 54.

The spark assumes very diverse aspects, according to the nature of the gas in which it is produced and the form of the conductors. Sometimes it takes the form of a straight, curved, or zigzag incandescent flash, and produces a violent detonation. A photograph of this spark shows that it is composed of a great number of interlaced luminous lines, marked with points more brilliant than the rest. At other times the spark seems to branch out, as shown in the bottom figure.

The duration of the spark from our machines is very short, for the illumination which it produces enables us to photograph objects in very rapid motion, such as projectiles; but the lightning-flashes which occur during storms often have sufficient duration to allow the shaking movement of leaves of trees to be distinguished at night.

Spectroscopic examination shows that the spark gives the lines of the metals composing the conductors, and of the gases in which it is produced, which gives us reason to attribute the light of the spark to the incandescence of particles violently torn away and volatilized. When the conductors are of different substances, it is observed that matter is transported from one to the other in both directions; a fact which gives evidence of an oscillating movement in the discharge-current. Further on will be seen a confirmation of this deduction.

When one of the conductors has sharp ridges or points, the discharge is continuous and takes the form of a phosphorescent, violet-tinted *brush*, accompanied by a peculiar hissing noise and a current of air called an electric breeze. Faraday has explained this current by the displacement of gaseous molecules, which are electrified by contact with the point and are then repulsed towards the conductors in the vicinity, which shows clearly the convective nature of the phenomenon. In rarefied gases the electric discharge is made evident by a glow attributed to the impact of the molecules which have been set in motion. When the vacuum is carried beyond a certain point, the glow only appears on the internal surface of the vessel containing the rarefied gas, which seems due to the fact that the gaseous molecules no longer impinge during their passage in the vessel, but strike against its walls. By modifying the nature of the glass of which the vessel is made we can give different colors to the glow.

The molecular agitation caused by the electric spark is especially favorable to chemical reactions; the combination of oxygen with hydrogen, the production of ozone, the odor of which accompanies the spark, are very well known phenomena. Ozone, which is used at the present time for the purification of alcohols and wines, is obtained by causing a current of air dried by sulphuric acid to pass through a space in which a glow-discharge is maintained.

Another property, shown by Prof. O. Lodge, is the condensation of vapors, smoke, or dust in a medium traversed by an electric spark. If electric discharges are produced in an atmosphere charged with particles in suspension, there arise vortices, followed by the precipitation of the particles on the surrounding walls. It has been proposed to apply this property to the condensation of the metallic dust in the flues of lead and zinc retorts, and even to the condensation of the heavy fogs which are formed in many cities.

LAWS OF THE ELECTRIC CURRENT.

111. Having now passed in review the different aspects of the discharge of electrified bodies, the law of the movement of electricity in conducting bodies will next be considered.

When the plates of a condenser are connected by a wire, the discharge is so rapid that the course of the phenomenon cannot be followed by any known experimental method. Its analysis is likewise extremely difficult in consequence of the effects of electromagnetic induction which are produced in the wire, as we shall see further on.

The complexity of the case of an electric flux in a conductor whose ends are at a variable difference of potential leads us to investigate first of all the case of a conductor whose ends are maintained at a constant difference of potential.

Let us first examine the means for reaching such a result.

112. Law of Successive Contacts. — We have seen, § 107, that two heterogeneous bodies furnish, on coming into contact, a difference of potential called *electromotive force of contact*, which is shown by means of an electrometer.

When a number of conductors, A, B, C, D, at the same temperature, form an open chain, the electromotive forces between the bodies in contact are added together or subtracted, according to their direction, and their algebraic sum constitutes the difference of potential between the extremities of the series. If these extremities be joined together so as to form a closed chain of conductors, the algebraic sum of the electromotive forces must be zero, or else a permanent current of electricity would be produced in the *circuit*, which would occur without any expenditure of energy.

Denoting by $A|B$, $B|C$, $C|D$ the successive differences of potential, we must have

$$A|B + B|C + C|D + D|A = 0,$$

whence

$$A|B + B|C + C|D = -D|A = A|D.$$

We see, then, that in a series of conductors connected one to another the difference of potential of the end conductors is the same as if they were directly connected.

The solder which joins two metals is therefore without effect upon their difference of potential.

113. Thermal and Chemical Electromotive Forces. Means of Keeping Up a Constant Difference of Potential in a Conductor. — The preceding law fails in two cases:

1. When the successive contacts are kept at different temperatures.

2. When the successive conductors act chemically upon each other.

The first case was discovered by Seebeck.

Suppose a copper wire forming a *closed circuit* with a zinc wire. When one of the junctions is heated, a flux or current of electricity is produced, flowing from the copper to the zinc across the hot junction. With the aid of a sufficiently delicate electrometer it will be found that in each of the wires the extremities are at a constant difference of potential, the temperature of these extremities remaining invariable.

The discovery of the second case is due to Volta.

Form a circuit by fastening a zinc and a copper wire together and plunging their free ends into dilute sulphuric acid, which *closes* the circuit. An electric current is produced in the conductors, for the electrometer shows, in the copper wire, potentials decreasing towards the zinc, and, in the zinc wire, potentials decreasing towards the acid.

These two experiments, which constitute the foundation of thermo-electric and hydro-electric batteries (to which we will return further on), furnish the means of maintaining a constant difference of potential between the ends of a conductive wire placed in the circuit. We will call these electromotive forces which form an exception to the law of successive contacts by the name of thermal or chemical electromotive forces.

114. Ohm's Law.—Suppose a homogeneous conductor, of any shape, two points of which, A and B, are maintained, as shown above, at different and constant potentials.

An electric field is formed in the conductor whose lines of force go from the point where the potential is highest, A, for example, towards the point with the lowest potential.

As the electricity can move along the lines of force, since the field is a conductor, it can be naturally admitted that in each tube of force there takes place a transportation of electricity proportional to the corresponding flux of force.

If, in an element of equipotential surface, the flux of force is $H\mathrm{d}s$, the flux of electricity per second will be represented by $\gamma H\mathrm{d}s$, γ being a constant for an isotropic conductor.

The total flux of electricity per second across an equipotential section of the conductor will be the sum of the elementary fluxes, such as $\gamma H\mathrm{d}s$.

$$I = \int \gamma H\mathrm{d}s.$$

This quantity is called the *strength of current*. The direction of the displacement of positive electricity is taken as the *direction of the current*. The factor γ is called the *specific conductivity*.

115. Case of a Conductor of Constant Section.— The ends of a homogeneous cylindrical conductor being maintained at constant potentials $U_1 > U_2$, the lines of force are parallel to the axis of the cylinder.

At any point

$$H = -\frac{\mathrm{d}U}{\mathrm{d}l},$$

$\mathrm{d}l$ denoting an element parallel to the axis.

As H is constant all along the cylinder, we have

$$H\int_0^l \mathrm{d}l = -\int_{U_1}^{U_2} \mathrm{d}U;$$

whence

$$Hl = U_1 - U_2.$$

Consequently the current

$$I = \int \gamma H\mathrm{d}s = \gamma \frac{U_1 - U_2}{l} \int \mathrm{d}s,$$

where the sign $\int$ must be extended to every section of the conductor; consequently

$$I = \frac{\gamma(U_1 - U_2)s}{l} = \frac{U_1 - U_2}{\dfrac{l}{\gamma s}}.$$

The same method of reasoning can be applied to a conductor of any shape and section whatever, provided that the latter be constant throughout the conductor.

The ratio $\dfrac{l}{\gamma s}$ is called the *resistance of the conductor ;* $\dfrac{1}{\gamma}$, which measures the resistance of a conductor having unit length and unit section, is the *specific resistance* or resistivity.

It is seen that the resistivity is the inverse of the conductivity, just as the resistance of a conductor is the inverse of its conductance.

The above equation was discovered by Ohm. It is enunciated as follows :

The strength of the current between two points of a conductor of any form, but of constant section, is directly proportional to the difference of potential and inversely proportional to the resistance between those points.

By virtue of the definition given of strength of current the quantity of electricity which passes across a section of the conductor in a time t is

$$q = It.$$

This evident equation is sometimes called *Faraday's law.*

116. Graphic Representation of Ohm's Law.—Ohm's law proves that in a conductor of constant section the potential falls uniformly, for the relation

$$dU = I\frac{dl}{\gamma s}$$

shows that the variation of the potential is proportional to the variation of the length.

Mark off on a horizontal line Ox a length OA, measuring, on a given scale, the resistance R of a conductor.

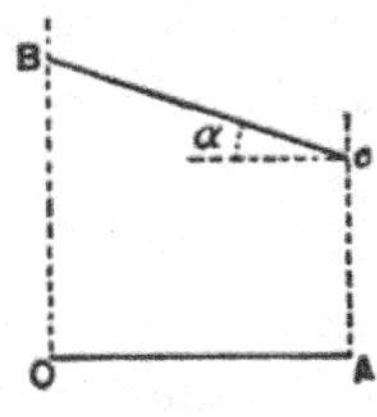

FIG. 55.

On the perpendiculars at the extremities of OA mark off OB and AC to represent the potentials at the two ends of the conductor; then the ordinates of the straight line BC represent the potentials of the intermediate points of the conductor.

The strength of the current is measured by the tangent of the angle between BC and OA, for we have

$$\tan \alpha = \frac{OB - AC}{OA} = \frac{U_1 - U_2}{R}.$$

117. Variable Period of the Current.—Suppose a conductor traversed by a current and enveloped in a dielectric sheathing, which latter is surrounded by a conductive layer: this is the case with an insulated conductor under water. If the outside conductor is connected with the earth, its potential is zero, and an *electric field* is formed in the dielectric, the lines of force passing from the inside conductor to the outside or *vice versa*, according as the potential of the inner conductor is greater or less than zero.

These lines of force join points carrying opposite charges (§ 88), so that the adjoining surfaces of the dielectric assume contrary electrifications, as in the case of a condenser.

The charge per unit length will, however, vary at different points of the conductor. If, for example, the potential decreases uniformly from U to o in the conductor, the charge will also decrease regularly from one extremity to the other, and the total charge of the conductor will be equal to the capacity of the condenser multiplied by the mean difference of potential of the two plates.

To explain the phenomenon of condensation we must admit that at the moment when a difference of potential occurs in the conductor a momentary flux of electricity traverses its surface to produce the surface-distribution. This *variable period* is followed by the *permanent* period, during which the electric flux is constant and passes in its entirety parallel to the walls of the conductor.

118. Application of Ohm's Law to the Variable Period of the Current in but Slightly Conductive Bodies.— When the plates of a condenser are joined by an imperfect insulator, such as caoutchouc or gutta-percha, it is observed that they lose their electrification little by little. This loss can be attributed to a certain conduction through the given substance, and Ohm's law can be applied to this special case of a variable period. Let C be the capacity of the condenser, R its resistance, U the difference of potential between the plates.

At any moment the current I is the ratio of the potential difference to the resistance,

$$I = \frac{U}{R}.$$

The quantity of electricity which passes during a time dt is

$$- \mathrm{d}q = I\mathrm{d}t$$

at that instant of time.

But

$$q = CU. \quad (\S\ 82.)$$

Consequently

$$dq = CdU;$$

whence

$$I = \frac{U}{R} = -\,C\frac{dU}{dt}.$$

The time that the difference of potential takes to pass from U_1 to U_2 is

$$\int_0^t dt = -\,CR\int_{U_1}^{U_2} \frac{dU}{U},$$

or

$$t = CR \log_e \frac{U_1}{U_2}.$$

119. Application of Ohm's Law to the Case of a Heterogeneous Circuit.—Suppose two conductors ab, bc, of resistances R_1 and R_2, in contact at the point b, and whose free ends a and c are maintained, by any means whatever, at constant potentials U_1 and U_2. A current will proceed from the point of higher potential to that of lower potential.

Suppose

$$U_1 > U_2.$$

At the point of contact b an electromotive force E is produced, being the difference of the potentials U_1', U_2', of the two sides of the surface of separation.

Suppose

$$U_1' > U_2',$$

we have

$$E = U_1' - U_2'.$$

Let us apply Ohm's law to the two portions ab, bc, observing that, as the electricity can be neither accumulated at nor abstracted from the point b during the permanent stage, the current must necessarily be the same in the two conductors.

We shall therefore have

$$I = \frac{U_1 - U'}{R_1} = \frac{U_2' - U_2}{R_2} = \frac{U_1 - U_2 - (U_1' - U_2')}{R_1 + R_2}$$

$$= \frac{U_1 - U_2 - E}{R_1 + R_2}.$$

In the case where

$$U_2' > U_1',$$

$$I = \frac{U_1 - U_2 + E}{R_1 + R_2},$$

the electromotive force E being taken with the $+$ or $-$ sign according as it gives an increase or decrease of potential in the direction of the current, that is to say, according as it tends to increase or diminish the current.

By extension, if there are several conductors in contact, E_1, E_2, E_3 ... being the electromotive forces, R_1, R_2, R_3 ... the resistances, we have

$$I = \frac{U_1 - U_2 + \Sigma E}{\Sigma R},$$

the signs of the electromotive forces being obtained by the preceding rule.

If we connect the end conductors directly, we will have

$$I = \frac{\Sigma E}{\Sigma R}.$$

We know that ΣE is zero, unless there is a difference of temperature or chemical action at the points of junction (§ 113).

120. Graphic Representation.—Take the case of three conductors; mark off successively on the axis of abscissæ lengths proportional to their resistances R_1, R_2, R_3.

Let $aa' = U_1$ (Fig. 56). From the point a' draw the right line $a'b'$ inclined at an angle α, such that

$$\tan \alpha = I.$$

If the electromotive force of contact E_1 is positive, measure it off along $b'b''$ and from b'' draw $b''c$ parallel to $a'b'$.

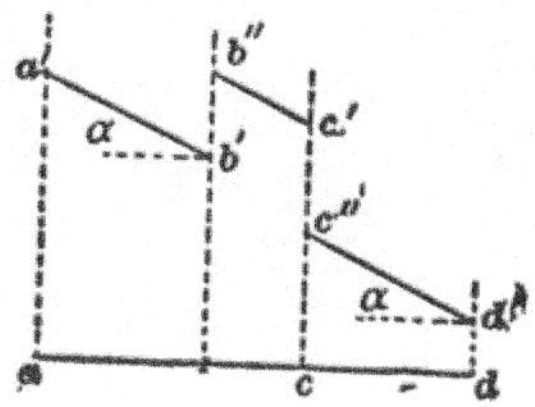

FIG. 56.

The electromotive force E_2, supposed to be negative, is marked off on $c'c''$ and the new line $c''d'$ will cut the potential-line dd' at the point d'.

121. Kirchhoff's Laws.—This name is given to two laws, one of which is evident when we compare the electric current with a fluid current, the other deduced from Ohm's law; the two together enabling us to solve the problem of even the most complicated electric circuits.

First Law.—When any number of conductors meet in a point, the algebraic sum of the currents at that point is zero.

This rule simply expresses the fact electricity can be neither accumulated nor subtracted at the meeting-point of conductors. The currents are considered as of opposite sign, according as they flow into or away from the point.

Second Law.—In every closed circuit, the algebraic sum of the electromotive forces equals the algebraic sum of the products of the currents by the resistances of the conductors.

Let $abcd$ be a circuit in a network of conductors.

Let us denote by i_1, i_2, i_3, i_4 the currents whose direc-

tion is shown by the arrows, and by r_1, r_2, r_3, r_4 the resistances, and by e_1, e_2, e_3, e_4 the electromotive forces shown
by the unequal parallel strokes. The $+$ sign shows the

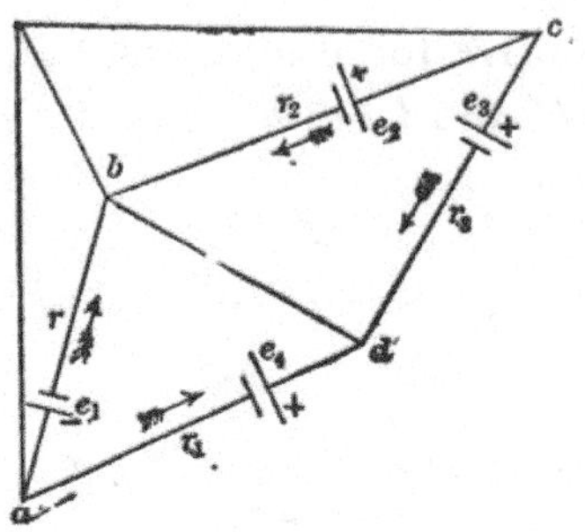

Fig. 57.

direetion in which each electromotive force tends to produce
an increase of potential.

Let u_1, u_2, u_3, u_4 be the potentials at the points a, b, c, d.
We shall then have, by § 113,

$$i_1 r_1 = u_1 - u_2 - e_1,$$
$$i_2 r_2 = u_3 - u_2 - e_2,$$
$$i_3 r_3 = u_3 - u_4 - e_3,$$
$$i_4 r_4 = u_1 - u_4 + e_4;$$

whence

$$i_1 r_1 - i_2 r_2 + i_3 r_3 - i_4 r_4 = - e_1 + e_2 - e_3 - e_4,$$

or

$$\Sigma i r = \Sigma e.$$

The signs which should be given to the currents and
electromotive forces are easily determined, viz., follow the
circuit round in the direction of movement of the hands of
a watch; give the sign $+$ to those currents which move
in this direction, and the sign $-$ to those moving in the
contrary direction. As to the electromotive forces, give
them $+$ or $-$ signs, according as they cause an increase or
diminution of potential in the given direction.

The application of Kirchhoff's laws to a combination of n conductors gives n distinct equations between the currents, the resistances, and the electromotive forces, whence we can deduce n of these quantities if the others are known.

This method will enable us, for example, to determine the currents and their signs: we begin by supposing arbitrary directions of current—the true directions will be given by the calculation ; a positive value will show that the supposed direction was correct, a negative value will indicate that the current was in the opposite direction.

122. Application to Derived Circuits.—The combination of circuits shown in Fig. 58 is made up of homogeneous conductors having resistances r_1, r_2, r_3, ending in the points a and b, which are connected by a conductor of the same nature, including a chemical source of electromotive force

Let r be the resistance of this part of the circuit.

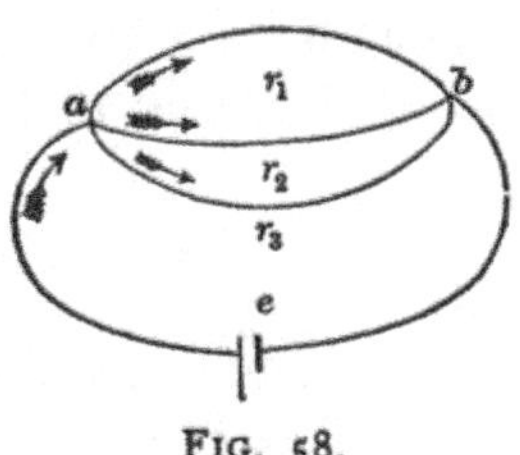

FIG. 58.

The electromotive force produces a total current I, which divides itself among the three *derived* branches r_1, r_2, r_3 into three partial currents i_1, i_2, i_3, such that

$$I = i_1 + i_2 + i_3. \qquad \text{(§ 111.)}$$

Kirchhoff's second law gives the equations

$$Ir + i_1 r_1 = e,$$
$$i_1 r_1 - i_2 r_2 = 0,$$
$$i_1 r_1 - i_3 r_3 = 0.$$

Eliminating successively i_1, i_2, i_3 from these four equations, we get

$$I = \cfrac{e}{r + \cfrac{1}{\cfrac{1}{r_1} + \cfrac{1}{r_2} + \cfrac{1}{r_3}}}.$$

The expression

$$\cfrac{1}{\cfrac{1}{r_1} + \cfrac{1}{r_2} + \cfrac{1}{r_3}}$$

represents the combined resistance of the three derived conductors r_1, r_2, and r_3.

In general, *the reciprocal of the combined resistance of a number of derived conductors is equal to the sum of the reciprocals of the resistances of the component conductors.*

123. Wheatstone's Bridge or Parallelogram.—The arrangement in Fig. 59 has been designed by Wheatstone for the purpose of measuring electrical resistances.

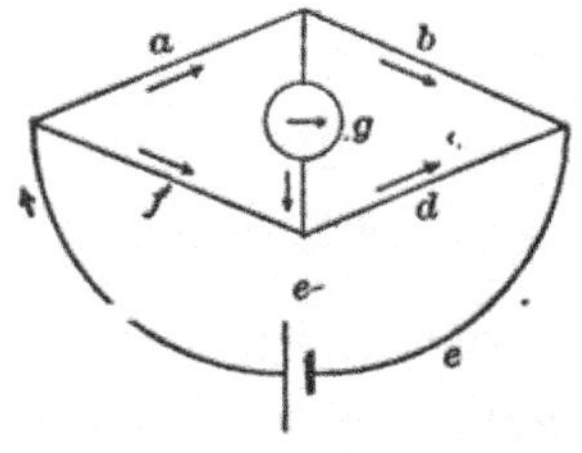

FIG. 59.

Suppose there are six conductors having resistances a, b, f, d, g, ρ. The branch r contains an electromotive force e; the branch g an apparatus to indicate the passage of a current.

The total current I is divided into partial currents which we will denote by the capitals A, B, F, D, G.

The application of Kirchhoff's laws gives the six following equations:

$$I - A - F = 0;$$
$$A - G - B = 0;$$
$$F + G - D = 0;$$
$$aA + gG - fF = 0;$$
$$bB - dD - gG = 0;$$
$$aA + bB + rI = e.$$

Eliminating A, B, F, D, we have

$$G = \frac{I(ad - bf)}{g(a + b + f + d) + (a + f)(b + d)}.$$

In order that the current G may be zero, it is sufficient
to have.

$$ad = bf \quad \text{or} \quad \frac{a}{b} = \frac{f}{d}$$

ENERGY OF THE ELECTRIC CURRENT.

124. General Expression.—In consequence of the definition of electric potential, when a quantity q of electricity
passes from a potential U_1 to a lower potential U_2, the work
accomplished is

$$(U_1 - U_2)q.$$

In the case where a current I circulates between the given
points the work per second, that is, the electric power developed by the current, is therefore

$$(U_1 - U_2)I.$$

If $U_1 - U_2$ represents an electromotive force E, thermic
or chemical, the power will be expressed by

$$EI.$$

125. Application to the Case of a Homogeneous Conductor. Joule's Effect.—If we take a homogeneous conductor of resistance R traversed by a constant current I, we have

$$(U_1 - U_2)I = I^2R. \qquad\qquad (\S\ 115.)$$

The work developed in a time t is

$$W = I^2Rt.$$

Joule has experimentally proved that this work is entirely transformed into heat in the conductor.

One of the most beautiful illustrations of Joule's effect is the incandescent electric lamp, in which the current heats a carbon filament placed in a glass bulb from which the air has been exhausted in order to prevent combustion.

126. Case of Heterogeneous Conductors. Peltier Effect.—Suppose a number of conductors R_1, R_2, R_3 ($\S\ 120$), without chemical action on each other; denote by I the current which flows through them, by E_1 and E_2 the electromotive forces of contact. By Joule's law the heat developed per second in each of the conductors is respectively

$$I^2R_1,\ I^2R_2,\ I^2R_3.$$

At the points of junction there are, in addition, abrupt variations of potential E_1, E_2 which correspond to amounts of electric energy E_1I, E_2I. The increase of energy of the current will be negative if the potentials fall in the direction of the current; it will be positive in the contrary case.

Peltier has shown, in the first case, that the junction is heated; in the second, that it is cooled. These calorific variations are equal and contrary to the variations in the energy of the electric flux. This phenomenon, known under the name of Peltier effect, enables us to measure exactly the electromotive force of contact. Contrary to the Joule effect, we see that the Peltier effect depends on the direction of the current and that it changes sign with the current.

To show the Peltier effect it is necessary to take steps to prevent the heat developed in the conductors by the Joule effect from hiding the variations of temperature, generally feeble at the points of junction. This is effected by using feeble currents and coating the junctions with some readily fusible substance, such as wax. The wax is then observed to melt when a current flows in one direction, and to solidify when it flows in the opposite direction.

From the law of successive contacts (§ 112) it follows that in a closed circuit where no differences of temperature are maintained by external sources of heat the algebraic sum of the electromotive forces of contact is zero, and consequently the sum of the Peltier effects is also zero.

127. Chemical Effect of the Current. Faraday's and Becquerel's Laws.—When an electric current passes through a compound liquid, by means of conductors or *electrodes* dipping into the liquid and kept at different potentials, besides the heating due to the Joule and Peltier effects, decomposition of the liquid is observed to take place.

The separated elements go to the electrodes, with which in certain cases they enter into combination.

This decomposition is called *electrolysis*, and the decomposed body the *electrolyte*. The electrode having the highest potential, by which the current enters, is the positive electrode or *anode;* the other is the negative electrode or *cathode.* The products of decomposition are the *ions.*

Electrolysis takes place according to the following (Faraday's) laws:

I. *The weights of the ions deposited and of the decomposed electrolyte are proportional to the quantities of electricity which have passed through the liquid.*

II. *When a number of electrolytes are traversed by the same current, the weights of the different ions set free are to each other as the chemical equivalents of these ions.*

The *electrochemical equivalent* of an ion or an electrolyte is the weight of this body deposited or decomposed per unit quantity of electricity.

Becquerel's Law.—In the case where two bodies form various combinations with each other the decomposition of these different combinations is dependent on the negative element. Thus in the electrolysis of the combinations PN_1, PN_2, P_2N_3, where P is a metal and N a metalloid, unit quantity of electricity sets free one electrochemical equivalent of N and weights of P equal to its electrochemical equivalent multiplied by 1, $\frac{1}{2}$, $\frac{2}{3}$.

128. Grothüss' Hypothesis.—The fact that the decomposition of an electrolyte is a necessity for the passage of the current has suggested the idea that the ions play the same part as the pith-ball in convective discharge.

If we admit that the molecules of the electrolyte are formed of groups of elements having opposite charges of electricity (possibly, according to Maxwell, due to the electromotive force of contact), then, at the moment of introducing electrodes, the positive elements or ions will turn towards the cathode and the negative elements towards the anode. This polarization will take place along the lines of force of the field produced in the liquid by the electrodes. If the intensity of the field is sufficient to overcome the chemical affinity of the compound, the ions near the electrodes are set free, while in the intermediate molecules of the liquid there is simply an exchange of elements. We have thus an explanation of the reason why the products of decomposition only make their appearance at the points where the current enters and leaves the liquid.

The electric charge carried per second by the positive ions to the cathode represents the amount of the current, which accounts for Faraday's first law. To account for the second law we need only suppose that the electronegative elements of different electrolytes have all the same electric charge.

According to Clausius' kinetic theory, the molecules are in motion and their impact causes their disassociation into the component atoms. But these atoms recombine with those liberated from the adjoining molecules, so that there are continual changes in all directions. The electric current causes an orientation of these movements and brings about a final decomposition at the electrodes.

129. Application of the Conservation of Energy to Electrolysis. Voltaic Cell.—The phenomenon of electrolysis can be considered from the point of view of the conservation of energy.

In an endothermic electrolytic reaction which absorbs energy, as is the case when acidulated water is decomposed between platinum electrodes, the effective energy of the current is diminished; a fall of potential takes place in the direction of the current, equal to what is called the *electromotive force of polarization* of the electrolyte. This electromotive force is negative (§ 119), and tends to produce a counter-current. The existence of this electromotive force can be shown by connecting the platinum conductors, immediately after electrolysis, to an apparatus for showing the passage of a current and then closing the circuit. A current will be observed directed from the cathode to the anode in the electrolyte, and at the same time the liberated elements, oxygen and hydrogen, will recombine.

Lord Kelvin has shown that the electromotive force of polarization can be calculated when the energy of combination of the electrolyte is known.   In fact, *if no secondary action takes place*, the electrical energy absorbed (represented by the product *ie* of the electromotive force of polarization into the current) is equal to the heat of combination of the weight of electrolyte decomposed per second, expressed in absolute units.   Let z be the electrochemical equivalent of the electrolyte, h the heat of combination of unit weight of the same.   We then have

$$ie = zhi\,;$$

whence

$$e = zh.$$

This expression gives the minimum difference of potential between the electrodes necessary to produce decomposition.

From these considerations we deduce a means of separating the elements of several electrolytes mixed together.   Suppose, for example, a solution containing sulphate of zinc and sulphate of copper ; as the heat of combination of the second salt is less than that of the first, we can, by suitably graduating the potential difference of the electrodes, deposit first copper and then zinc upon the cathode.

There are cases where the ions react on the electrodes, giving rise to new compounds.   The energy set free in these reactions must be taken into account in calculating the electromotive force necessary for decomposition.

Take the example of the electrolysis of a solution of copper sulphate between copper electrodes.   The liberated copper will go to the cathode and the acid to the anode. which it will dissolve equivalent for equivalent.   The reaction at the electrodes thus neutralizes the chemical effect of the current, so that the electromotive force of decomposition is zero.   The whole energy of the current is used in the Joule effect, that is, in heating the bath.

Suppose that water acidulated with sulphuric acid is decomposed between a zinc anode and a copper cathode. The hydrogen will be deposited on the copper, while the oxygen will form oxide of zinc with the anode, which will then be dissolved in the state of sulphate of zinc. But as the heat of combination of zinc sulphate is higher than that of sulphuric acid, a quantity of energy is consequently liberated which shows itself by an increase of potential in the direction of the current equal to the effective difference of the electromotive forces.

This electromotive force is $E = ph - p'h'$, ph denoting the heat of formation of one electrochemical equivalent of zinc sulphate, $p'h'$ that of one equivalent of sulphuric acid.

Such a combination, called *voltaic element* or *couple*, is a source of electricity; and if we connect the two electrodes by a copper wire it will be found that it is traversed by a current flowing from the zinc to the copper in the electrolyte, and from the copper to the zinc in the *external circuit* formed by the wire.

The direction of the current in the external circuit shows that the copper is at a higher potential than the zinc, whence the name of *positive pole* or *plate* given to the plate of copper. The zinc plate is, conversely, called the *negative pole* or *plate*.

The current is

$$I = \frac{E}{R},$$

where E represents the electromotive force and R the resistance of the circuit, including the resistance of the liquid as well as that of the wire and the electrodes.

THERMO-ELECTRIC COUPLES.

130. Seebeck and Peltier Effects.—Seebeck has proved (§ 113) the production of an E. M. F. in a chain of metals whose junctions are maintained at unequal temperatures. Thus on forming a circuit of an iron and a copper wire the latter including a galvanometer coil, and on raising the temperature of one of the points of junction, a current is set up in the system going from the copper to the iron across the hot junction. Such a circuit is called a *thermo-electric couple.*

The Seebeck effect is reversible, as has been shown by Peltier; when a current due to an outside E. M. F., traverses the junction of the two metals, from the copper to the iron, the junction is cooled; if the direction of the current is reversed, it grows hot. This phenomenon is distinct from the Joule effect; but, as the two effects occur simultaneously, certain precautions must be taken in order to distinguish one from the other.

The development of heat due to the Peltier effect is proportional to the first power of the current, while the Joule effect depends on its square; it is therefore advantageous to employ feeble currents in order to distinguish the two actions.

According to Maxwell, the Peltier effect is a measure of the E. M. F. of contact, § 107. In fact the heat developed at the junction by a current i in one second is expressed by the product ei, in which e represents the difference of potential which is set up on the contact of the given bodies. If this heat n is given in gramme-degrees, e in volts, and i in ampères, we have $4.2n = ie$; whence

$$e = \frac{4.2n}{i} \text{ volts.}$$

The values found by this means depend on the absolute temperature of the junction $e = f(T)$; they are very slight compared with the potential differences observed on connecting to the terminals of an electrometer two points of the conductors taken on each side of the junction and close to it.

Thus the E. M. F. of contact of zinc and copper measured by the first method is, at 25° C., 0.00045 volt, while the electrometer indicates about 0.8 volt; but Maxwell has observed that, in the latter case, we have not only to do with the contact $Zn \mid Cu$, but that these metals form, with the air which separates the fixed parts from the movable parts of the electrometer, a closed chain:

$$Zn \mid Cu + Cu \mid \text{air} + \text{air} \mid Zn.$$

Now the E. M. F. of contact of the air with the metallic parts of the instrument, by virtue of which they assume electric charges, may be much greater than that which is developed in an entirely metallic contact, which would explain the observed anomaly.

In the thermo-electric series a metal is said to be positive with regard to another when the E. M. F. of contact is directed from the first to the second across the heated junction.

The E. M. F. which is set up in a metallic arc formed of two dissimilar metals depends, as might be expected, on the heat communicated to one of the junctions, and the simplest way is to express this E. M. F. as a function of the difference between the temperatures θ, θ' of the two junctions of the metals forming the circuit. But we have seen that this E. M. F. is also dependent on the absolute temperature of the two junctions, or, in other words, on the mean of their temperatures:

$$e = f\left(\frac{\theta + \theta'}{2},\ \theta - \theta'\right).$$

Thus, when, in the case of the above-mentioned couple, we make one of the junctions continually hotter while keeping the other at a constant temperature, the thermo-electric current increases to a maximum, then decreases, becomes zero, and ends by changing direction.

131. Kelvin Effect.—In seeking for the cause of this peculiarity Lord Kelvin discovered that the E. M. F. which gives rise to the current is not situated at the junctions alone, as was long believed, but that the homogeneous wires, unequally heated, which form the circuit are also the seat of electromotive forces.

Thus in a metal bar an unequal distribution of temperature causes differences of potential between the various points. If the temperature rises from one end of the bar to the other, a continual rise of potential is observed for some metals, while others give a fall of potential in the direction corresponding to the rise of temperature.

According to M. Leroux, lead is the only metal in which similar electric phenomena are not manifested when various parts of a piece of this substance are put in different thermal conditions.

These E. M. F.'s combine with those which are set up at the points of junction and give a resultant E. M. F., whose ratio to the resistance of the circuit is the strength of the thermo-electric current. If the sum of the potential differences set up in the metals is opposite to the sum of the potential differences at the junctions, it may be assumed that there exist temperatures for which these sums are equal, and the thermo-electric current ceases. The mean of the temperatures of the junctions for which this phenomenon occurs is called the *neutral temperature* or *temperature of reversal.*

The properties discovered by Lord Kelvin, and which are

called the *Kelvin effect*, are, from a certain point of view, reversible.

Thus, when a current is passed through a wire whose ends are maintained at different temperatures, and which consequently has a distribution of potential of its own, the current cools the wire if it is directed towards increasing potentials, and heats it in the contrary case.

In order to exhibit this effect independently of the Joule effect proceed as follows: A metal bar is heated towards the middle, while its extremities are kept at 0° in melting ice. When a current passes in the bar, it is found that points situated symmetrically with respect to the middle are not at the same temperature; in fact there is a greater heating in one of the halves where the Joule and Kelvin effects are added together; in the other half the Kelvin effect absorbs part of the heat due to the Joule effect. The action is the same as if there were a transference of heat in the bar; the transfer is made in the direction of the current for certain metals, and in the opposite direction for others. Lead is the only metal which preserves a perfect symmetry as regards the distribution of the temperature.

It is interesting to observe that, in a thermo-electric chain whose junctions are maintained at unequal temperatures, the algebraic sum of the potential differences is zero, as indeed is the case in every closed electric circuit. It follows that the heating observed at the points where the current experiences a drop in potential exactly compensates the corresponding cooling at the rises of potential; in other words, the total heat produced in the circuit by the Peltier and Kelvin effects is zero. This is not the case with the heat corresponding to the Joule effect, whose value, necessarily positive, represents the quantity of heat supplied at the hot junction, excepting for losses by radiation and conductivity.

132. Laws of Thermo-electric Action.—The two following laws were discovered experimentally by Becquerel:

Law of Successive Temperatures.—In a thermo-electric couple formed of two dissimilar bodies the E. M. F. corresponding to two temperatures θ_1 and θ_2 of the junctions is equal to the algebraic sum of the E. M. F.'s corresponding to the temperatures θ_1 and θ on the one hand, and θ and θ_2 on the other.

Law of Intermediary Metals.—If two metals in a circuit are separated by one or more intermediary metals, all kept at the same temperature, the E. M. F. is the same as if the two metals were directly united and their junction raised to the same temperature ; consequently the solder placed between two metals is without effect on the E. M. F. of the couple.

These laws are supplemented by those of Kelvin and Tait, which may be announced as follows :

Kelvin's Law.—If the extremities of a homogeneous bar are kept at temperatures θ and θ', an E. M. F. exists in the bar proportional to $\theta' - \theta$, the coefficient of proportionality, itself variable with the temperature, is called by Kelvin the *specific heat of electricity.*

These laws being granted, let us consider a couple formed of the metals A and B, whose junctions are maintained at temperatures θ and θ'; let U and U' be the sudden variation of potential at junctions, σ and σ' the specific heats of electricity of A and B. The total E. M. F. will be

$$e = U - U' + \int_{\theta}^{\theta'} (\sigma - \sigma') d\theta.$$

Tait's Law.—The specific heat of electricity of a body is proportional to its absolute temperature, $\sigma = K\theta$.

133. Thermo-electric Powers.—The E. M. F.'s which are set up in thermo-electric couples by the effect of progressive variations of temperature are determined by observing the currents which result from them in circuits of known resistance. The increase of resistance caused by the rise of temperature of one of the junctions is rendered negligible by introducing into the circuit a large supplementary resistance, which may be of the same substance as one of the metals in the couple, or of a different substance on condition that it is kept at the same temperature in all its points (law of intermediary metals).

Suppose that one of the junctions be kept at an invariable temperature by immersion in melting ice, for example, and the other junction raised to increasing temperatures by immersion in a heated bath containing a thermometer. The difference of temperature and the total E. M. F.'s are to be observed simultaneously. To represent the phenomenon graphically the values of the former may be made abscissæ and those of the latter ordinates; the curves thus drawn are very sensibly parabolas with vertical axes, their apex corresponding to the temperature of reversal (Gaugain).

Their equation is of the form

$$e = k + a\theta + \frac{b\theta^2}{2}.$$

From this we deduce

$$\left(\frac{de}{d\theta}\right)_\theta = a + b\theta.$$

This derivative, which is the angular coefficient of the tangent to the parabola, is the E. M. F. corresponding to a difference of temperature of $1°$ between the junctions at the mean temperature θ. This value has by Lord Kelvin been given the name of *thermo-electric power* of the couple at the given temperature.

At the neutral temperature the tangent is parallel to the axis of the abscissæ, whence

$$\left(\frac{de}{d\theta}\right)_n = 0 = a + b\theta_n, \text{ and } \theta_n = -\frac{a}{b}.$$

The total E. M. F. corresponding to temperatures θ and θ' at the junctions is given by the knowledge of the coefficients a and b:

$$e = \int_\theta^{\theta'} (a + b\theta)d\theta = a(\theta - \theta') + \frac{b}{2}(\theta^2 - \theta'^2)$$

$$= b(\theta' - \theta)\left[\theta_n - \frac{\theta + \theta'}{2}\right].$$

This formula shows that, when the temperatures θ and θ' are equidistant from t_n, the E. M. F. is zero.

In order to determine the pairs of parameters a and b or b and t_n we need only perform experiments at temperatures t and t', t_1 and t_1'; we thus get two equations in which the quantities sought are the only unknown ones:

$$e = a(\theta - \theta') + \frac{b}{2}(\theta^2 - \theta'^2);$$

$$e' = a(\theta_1 - \theta_1') + \frac{b}{2}(\theta_1^2 - \theta_1'^2).$$

In order to represent graphically the variations of the thermo-electric power of a couple $A \mid B$ with the temperature, we need only draw the right line MM', whose equation is

$$\frac{de}{dt} = a + b\theta.$$

For another couple $A \mid C$ we shall have a second right line NN' usually cutting the first.

Now by the law of intermediary metals the thermo-electric power of the couple $C \mid B$ will be given by the difference of the ordinates of MM' and NN'.

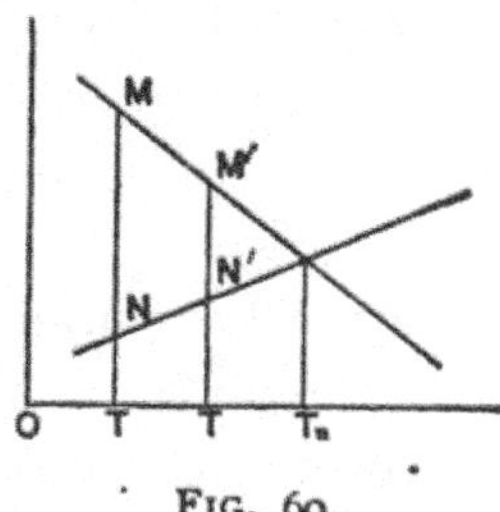

FIG. 60.

Knowing the parameters of the two couples $A \mid B$ and $A \mid C$,

$$\left(\frac{\mathrm{d}e}{\mathrm{d}\theta}\right)_{A \mid B} = a + b\theta,$$

$$\left(\frac{\mathrm{d}e}{\mathrm{d}\theta}\right)_{A \mid C} = a' + b_1\theta,$$

we deduce those of the couple $C \mid B$ from the formula

$$\left(\frac{\mathrm{d}e}{\mathrm{d}\theta}\right)_{A \mid B} - \left(\frac{\mathrm{d}e}{\mathrm{d}\theta}\right)_{A \mid C} = \left(\frac{\mathrm{d}e}{\mathrm{d}\theta}\right)_{C \mid B} = a - a' + (b - b')\theta.$$

We need therefore only draw diagrams of the thermo-electric powers of all the metals taken separately with one of their number in order to learn the values of the thermo-electric powers of all the metals taken in pairs in any combination.

Lead is generally adopted as the metal of comparison, because its specific heat of electricity is zero.

In these diagrams the intersection of two right lines has as its abscissa the value of the temperature of reversal.

It will be observed that the E. M. F. of a couple $A \mid B$ between two temperatures θ and θ' is expressed by

$$\int_\theta^{\theta'} \frac{\mathrm{d}e}{\mathrm{d}\theta}\mathrm{d}\theta = \int_\theta^{\theta'} (a + b\theta)\mathrm{d}\theta,$$

being represented by the area included between the right line *MM'*, the axis of abscissas and the extreme ordinates corresponding to θ and θ'. Likewise the E. M. F. of $C \mid B$, between the same temperatures, is represented by the area *MM'N'N.* This area can also represent the work done by a quantity of electricity equal to one coulomb traversing the circuit $C \mid B$.

We give here the values of the parameters a and b, which enable the thermo-electric powers of various bodies, taken with lead, to be calculated in microvolts:

	a	b
Copper	— 1.34	— 0.0094
Alloy (90 Pt + 10 Ir)	— 5.90	+ 1.0133
Iron	— 17.15	+ 0.0482
German-silver	+ 11.94	+ 0.0506

From these figures it is seen that an iron-german-silver couple has a thermo-electric power of

$$(- 29.09 - 0.0024\theta) \text{ microvolts.}$$

The current goes from the german-silver to the iron across the hot junction. The E. M. F. for the temperatures 0° and 200° at the junctions is 5.866 millivolts.

These results show that thermo-electric couples only produce extremely feeble E. M. F's, and that it is consequently necessary to join up a large number of them in series in order to obtain differences of potential comparable to those obtained in hydro-electric batteries. It is true that, as the couples are formed of very good conductors, they are capable of producing tolerably strong currents in an external circuit of small resistance.

Various bodies give E. M. F.'s very much higher than those of the common metals, but they cannot stand as high temperatures. According to Becquerel, at a temperature of

50° C. the thermo-electric power of the bismuth-lead couple
is $+$ 40 microvolts and that of the fused copper-sulphate-
lead couple is $-$ 352 microvolts. Antimony-zinc alloy (in
equal proportions) gives with lead $-$ 98 microvolts. The
conductivity of these different bodies is very inferior to that
of the metals, and it is necessary to use them in the form of
tolerably thick bars.

134. Thermo-electric Pile.—To form a thermo-electric
pile a chain is made in which the alternate links are formed
of one of the two metals chosen to form a couple, and all
the odd (or all the even) junctions are heated.

In order not to need to use as many heaters as there are
couples the chain is folded zigzag, the consecutive links being isolated with asbestos; in this way a solid block is obtained, with the even junctions on one side and the odd ones
on the opposite side ; then only one source of heat is needed
to heat all the junctions on one side. The opposite junctions
can be cooled by a current of air ; they are also often furnished with expansions intended to aid the radiation of the
heat transmitted across the couples by conductivity ; for
this purpose are used thin sheets of copper or iron, blackened
in order to increase their emissive power.

It frequently happens that the substances used to make
the couples are not very capable of supporting the direct
action of the flames which heat the junctions. In this case
the latter are covered with a solid envelope upon which the
flame plays, and which transmits the heat to the couples by
conductivity. This arrangement has also the advantage of
rendering the variations of temperature in the couples less
sudden when the fire is lighted or extinguished, and consequently of diminishing the disintegration which takes
place, in consequence of these sudden changes, in the bars
of alloy employed.

ELECTROMAGNETISM.

MAGNETIC PHENOMENA DUE TO CURRENTS.

135. Oersted's Discovery. — Oersted established in 1820 the action of an electric current upon a magnet-needle. This discovery was the point of departure for the theory of electromagnetism, which was established almost entirely by Ampère. According to the practical rule pointed out by this physicist, the north pole of the needle

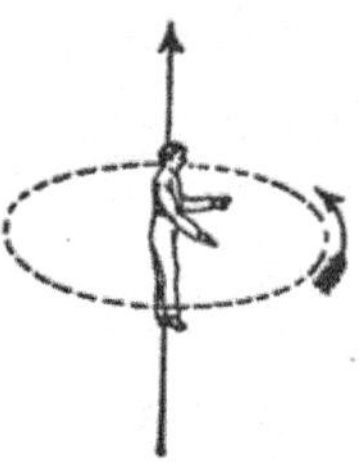

FIG. 61.

tends to move towards the left hand of an observer who looks at the needle when he is placed in the direction of the current, so that it enters at his feet (Fig. 61). This fundamental action shows that the current produces a magnetic field, which fact can be also shown by the use of iron-filings (§ 47).

Upon dusting iron-filings on a sheet of paper traversed by a current perpendicular to the plane of the paper it is seen that the particles form circles whose centre is in the axis of the conductor. A magnetic pole free to move

184

about the conductor would consequently tend to turn around it. The direction of this movement can be determined by Ampère's rule, or by that of Maxwell, which is often more convenient to apply. The direction of rotation

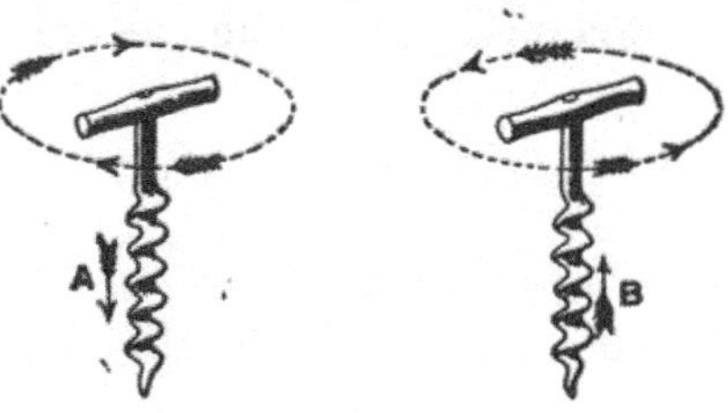

FIG. 62.

of the north pole and the direction of the current are indicated by the relative movements of rotation and translation of a corkscrew.

The circular form of the magnetic lines of force due to a rectilinear current explains why a magnet-needle tends to place itself transversely to the current, so that its magnetic axis may be tangent to the line of force which passes through its support.

136. Magnetic Field due to an Indefinite Rectilinear Current.—The intensity at different points of the field may be studied by the method of oscillations (§ 47). By applying this method, Biot and Savart have found that the intensity of the field due to a rectilinear current, sufficiently long and distant from the rest of the circuit to be considered as indefinite, is proportional to the current and inversely proportional to the distance from the conductor. The direction of the field is normal to the plane passing through the conductor and the point under consideration. The force exercised on a positive pole m can therefore be expressed by

$$F = \frac{kim}{l}.$$

The intensity of the field at a distance l is consequently

$$\mathcal{H} = \frac{ki}{l}.$$

As the reaction is equal and contrary in direction to the action, a pole m exercises upon the current a force equal to

$$\frac{kim}{l}.$$

This force is directed towards the *right-hand* of Ampère's manikin when he faces the pole, or towards his *left* if he is looking in the direction of the lines of force emerging from the pole.

137. Laplace's Law.—Biot has investigated the action of a current traversing two indefinite rectilinear conductors, AB, AC, placed at an angle, upon a pole m situated on the

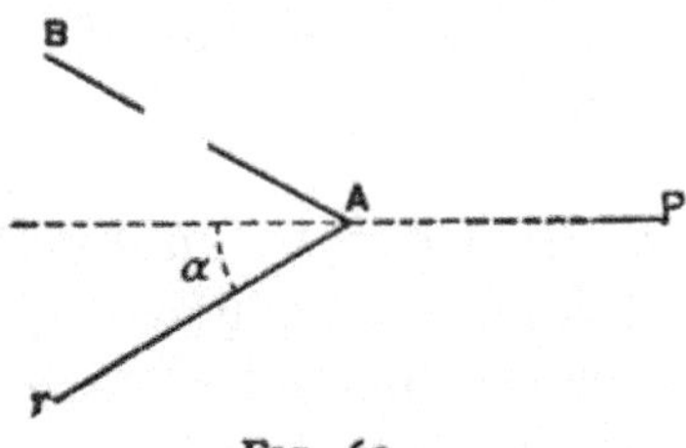

Fig. 63.

line bisecting the angle (Fig. 63). He found that the force can be represented by

$$F = k\frac{im}{l} \tan \frac{1}{2}\alpha,$$

k being a coefficient of proportion, l the distance from the pole P to the apex A, α the half-angle between the conductors. This expression reduces to that of the preceding paragraph when $\alpha = 90°$.

The direction of the force is normal to the plane of the two conductors. Laplace has deduced from this expression the action of an element of current on a pole.*

By reason of symmetry, the effect of one of the branches AB is

$$F = \frac{k}{2}\frac{im}{l}\tan \tfrac{1}{2}\alpha = k'\frac{im}{l}\tan \tfrac{1}{2}\alpha, \quad . \quad . \quad . \quad (1)$$

α being the angle made by AB with PA.

Prolong the branch BA by a quantity $AA' = ds$, and find the action of this element of current on the pole situated at P.

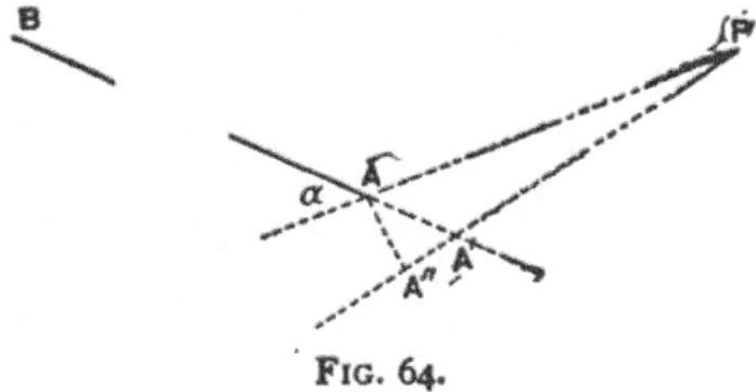

Fig. 64.

It will be observed that F is a function of two variables, r and α, which determine the relative positions of m and ds. We can then write the identity

$$dF = \frac{dF}{ds}\,ds = \left(\frac{dF}{d\alpha}\frac{d\alpha}{ds} + \frac{dF}{dr}\frac{dr}{ds}\right)ds. \quad . \quad . \quad . \quad (2)$$

In order to obtain the expression for df, we need only substitute in (2) the values of the four derivatives, deducing

---

* The following demonstration is due to M. de Weydlich, formerly assistant at the Liège Electrotechnical Institute.

them from the results of experiments and from geometrical considerations. Drawing from the point P as centre the arc AA'', and observing that the angle $APA'' = d\alpha$, since it is the increase of the angle between the directions of AB and PA, we will have in the infinitely small triangle $AA'A''$

$$AA'' = ds \sin \alpha = l d\alpha,$$

whence

$$\frac{d\alpha}{ds} = \frac{\sin \alpha}{l}$$

and

$$A'A'' = - dl = ds \cos \alpha,$$

whence

$$\frac{dl}{ds} = - \cos \alpha.$$

On the other hand, equation (1) gives directly

$$\frac{dF}{d\alpha} = \tfrac{1}{2} \frac{k'im}{l} \frac{1}{\cos^2 \frac{\alpha}{2}}.$$

$$\frac{dF}{dl} = - \frac{k'im}{l^2} \tan \frac{\alpha}{2}.$$

Substituting these expressions in (2), we find

$$dF = \frac{k'im}{l^2} \left(\tfrac{1}{2} \frac{1}{\cos^2 \frac{\alpha}{2}} \sin \alpha + \tan \frac{\alpha}{2} \cos \alpha \right) ds$$

$$= \frac{k'im}{l^2} \tan \frac{\alpha}{2} (1 + \cos \alpha) ds$$

$$= \frac{k'im}{l^2} \sin \alpha ds = \frac{k'im}{l^2} ds \sin (l, ds).$$

The elementary force is normal to the plane of the current and of the pole. If we consider the reaction of the pole on the element ds, we find it directed towards the right hand of Ampère's manikin when placed in the direction of the current and facing the pole (§ 136), or towards his left if turned so as to face in the direction of the lines of force produced by the pole.

138. Action of a Magnetic Field on an Element of Current.—It will be noticed that in Laplace's law the factor $\dfrac{m}{l_2}$ represents the intensity of the field $\mathfrak{H}$ due to the pole m, at the point where the current-element is situated. We can therefore write

$$dF = ki\,\mathfrak{H}\,ds\,\sin(\mathfrak{H},\,ds).$$

It is easy to generalize the law for the case of a number of poles.

The total force dF is the product of $kids$ by the resultant of the terms such as

$$\frac{m}{r_2}\,\sin(l,\,ds),$$

which represents the product of the intensity of the field $\mathfrak{H}$ by the sine of the angle between the direction of the field and the direction of the element, since the projection of the resultant is equal to the sum of the projections of the components.

Consequently the resultant is, in absolute value,

$$dF = ki\,ds\,\mathfrak{H}\,\sin(\mathfrak{H},\,ds).$$

From what we have learned in the preceding paragraph, this force, applied to an element of current, is normal to the plane of the current and the field, and directed towards the *left* of Ampère's manikin when he faces in the direction of the lines of force of the field.

Fleming has pointed out another way of determining the direction of the electromagnetic force. If the thumb and first two fingers of the left hand be pointed in three directions perpendicular to each other, pointing the fore and middle fingers respectively in the direction of the magnetic lines of force and the current, then the thumb points in the direction in which this last tends to be displaced.

139. Work due to the Displacement of an Element of Current under the Action of a Pole.—Suppose a current-element ds $= ab$ (Fig. 65) acted on by a pole situated in a point P.

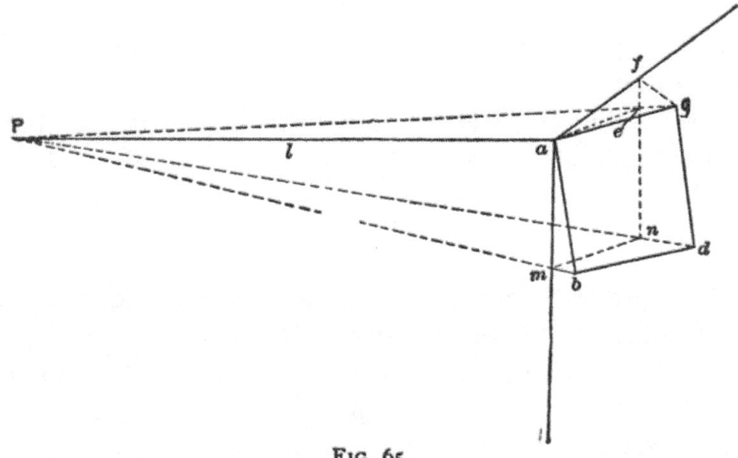

FIG. 65.

The electromagnetic force tends to displace the element ds perpendicularly to the plane l, ds following a direction af. If the element moves in a direction ag, the work performed during a displacement ds' $= ag$ is equal to the product of the force by the projection of the displacement along the direction of the force.

We then have

$$dW = \frac{kim}{l^2}\, ds\, \sin\,(Pab)\, ds'\, \cos\,(gaf). \quad . \quad . \quad (1)$$

Now construct a parallelogram on ag and ab; pass through af a plane fam normal to l and determine the intersections e, n of this plane with the right lines gP, dP.

As ag and ab are infinitely small in comparison with r, the plane gdP is normal to fam and contains the right line gf which is projected in ag on af.

Equation (1) can consequently be written

$$dW = \frac{kim}{l^2} \times \overline{am} \times \overline{af}.$$

But the product $\overline{am} \times \overline{af}$ measures the surface of the parallelogram $aenm$, of which af is the altitude, this parallelogram being capable of being considered as the projection of $agdb$ on a sphere of radius l and centre P.

Dividing this projection by l^2, we obtain the projection on a sphere of unit radius, or the solid angle subtended by the parallelogram $agdb$, that is to say, the area described by the element ds. Calling this solid angle $d\omega$,

$$dW = kimd\omega.$$

140. Work Due to the Displacement of a Circuit under the Action of a Pole.—To find the work performed by a current of finite length displaced under the action of a pole, we need only consider the sum of such terms as $kimd\omega$. We find the product of kim by the solid angle subtended at the pole by the surface described by the given current.

In the case of a closed circuit, $abgd$, Fig. 66, which assumes a position $a'b'g'd'$ under the action of a pole situated in the direction of the reader, we can consider separately the segments *abg*, *gda* traversed by opposite currents. The work accomplished by *abg* is proportional to the solid angle

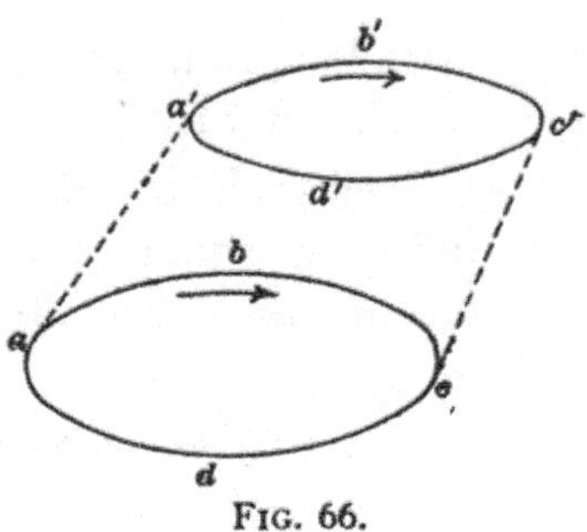

FIG. 66.

subtended by the area *aa′b′g′gb*; that by *adg* is proportional to the apparent surface of *aa′d′g′gd*. The resultant work will be proportional to the difference between these apparent surfaces, that is, to the difference between the solid angles subtended by the two contours *a′b′g′d′*, *abgd* of the circuit.

It follows from the preceding, that to bring a pole *m* from an infinite distance to a point at which the outline of the circuit subtends an angle ω, the work performed is

$$-kim\omega,$$

ω being the solid angle subtended by that face of the current which attracts a positive pole.

This expression therefore represents the relative energy of the current and the pole. If the latter were a unit pole, the work would be $-ki\omega$.

141. Magnetic Potential Due to a Circuit. Unit of Current. Ampère's Hypothesis on the Nature of Magnetism.—It will be observed that the expression $-ki\omega$ answers to the definition of potential in terms of work (§ 12).

We are moreover justified to define the magnetic forces
due to the current by a potential, since the work accomplished in the field of the current is a function of the coordinates of the circuit traversed by the current, and of the
point where the magnetic pole is supposed to be placed.
The expression $-ki\omega$ can therefore be called the magnetic
potential due to the current at the given point where the
unit pole is placed :

$$U = -ki\omega. \tag{1}$$

On comparing the potential $-ki\omega$ due to the current
with the potential $-F_i\omega$ due to a magnetic shell (§ 43), we
recognize the identity of the two expressions. A current
gives the same potential, and consequently produces the
same magnetic forces, as a shell of the same outline, whose
strength F_i would be equal to ki. To determine the direction of the forces and the sign of the potential, it is necessary to distinguish the two faces of the circuit. According
to Ampère's rule (§ 135), the face of the circuit which produces the same action as the negative side of a shell (that
is, attracts a north pole) is that around which the current
seems to go in the direction of the hands of a watch : this is
the negative or S face of the current, the other is the positive or N face.

The numerical coefficient k of equation (1) depends on the
unit chosen to measure the current. It could be put equal
to unity, and the unit current defined as that current which
produces unit magnetic potential at a point where the circuit subtends unit solid angle. The unit thus chosen has
the same dimensions as the unit of strength of a shell

$$[L^{\frac{1}{2}}M^{\frac{1}{2}}T^{-1}].$$

This is the unit which we shall adopt in future to express the

current ; we will later on come across more tangible definitions of it.

Ampère, who discovered the identity of effect of a shell
and a current, interpreted this identity in the following
manner : Experiment shows that a small closed current acts
like a small magnet normal to the plane of the current on
condition that the moment of the magnet be equal to the
current multiplied by the surface of the circuit. Suppose
any finite circuit divided by lines drawn across it interiorly
in the form of a network with an infinite number of meshes,
and suppose that the edges of these meshes are traversed
by equal currents in the same direction. The result will be
currents in the interior lines of the network which annul
each other in pairs; the outer contour of the circuit will be
the only seat of a current. Replacing each mesh by an
equivalent elementary magnet, the combination of these elements forms a magnetic shell whose effect is identical with
that of the current having the same contour.

From the preceding comparison Ampère deduced an hypothesis which refers magnetic phenomena to electric
phenomena. It need only be admitted that each atom of a
magnet is the seat of a circular current; the orientation of
these currents will produce effects identical with those of
magnets. We must then adopt as a postulate that such an
elementary current can exist without expenditure of work,
that is, that electric resistance is only manifested in traversing interatomic spaces.*

Fleming and Dewar have recently advanced an experimental proof in favor ofth is latter hypothesis. They showed
that if the variations of the resistance of pure metals are
represented graphically in terms of their temperatures,
curves are obtained which appear to converge toward a
point of no resistance, or absolute zero.

---

* See Ampère, *Mémoires publiés par la Société de Physique.*

There is, however, a distinction to be established between the potential due to a current and that given by a shell. Suppose a positive magnetic mass equal to unity be placed against the positive face of a shell. It will be repelled, will follow a curved trajectory, called line of force, and will bring up against the negative face, where it will remain in stable equilibrium. The work accomplished during this revolution is $4\pi F$, (§ 44).

In the case of a current, the N pole will also follow a line of force, but as this latter is a continuous curve, the pole will continue to move in this orbit as long as the current lasts. Each revolution will increase the work done by the circuit by $4\pi i$, and consequently, by definition, the potential will be expressed by

$$U = - i(\omega \pm 4\pi n),$$

n denoting the number of revolutions described by the unit pole. If the work has been performed by the magnetic force due to the current, we must take the sign $+$, for the potential will then have decreased; in the contrary case, that is when the pole has been forced to move backwards, the sign $-$ must be chosen.

It follows from the preceding that the magnetic potential due to a current contains one constant more than the potential of a shell. Nevertheless, as regards the determination of the forces, this constant is eliminated, since the intensity of the current's magnetic field is, in a direction l,

$$\mathcal{H} = \frac{dU}{dl} = + i\frac{d\omega}{dl}.$$

We can therefore say that as regards exterior magnetic actions a current i is comparable to a shell F, having the same contour.

The total magnetic flux produced by the current is the sum of the terms of the same sign, such as $\mathcal{H}ds$, which can be formed in an equipotential surface.

The electromagnetic action of a current can be summed up by saying that the current magnetizes the medium which surrounds it, and develops a flux of magnetic force proportional to the intensity of the electric flux and the permeability of the medium. A current surrounded with iron will produce a very much greater flux than if it were surrounded by air or some feebly magnetic substance.

142. Energy of a Current in a Magnetic Field. Maxwell's Rule.

—Following up the comparison between shells and electric circuits, and extending the expression $-im\omega$ which has been found for the relative energy of a current and a pole, it is easy to see that the relative energy of a current and a field is

$$W = -i\Phi,$$

Φ denoting the flux of force across the negative face of the circuit.

If the current is displaced in the field, the work accomplished is measured by the variation of potential energy. When this latter becomes a minimum the circuit reaches a position of stable equilibrium, which corresponds to a maximum flux of force penetrating across the negative face of the current. Hence Maxwell's rule:

A current free to move in a magnetic field tends to place itself so as to receive the greatest possible flux of force across its negative face.

Thus, a circular current movable about one of its diameters which is perpendicular to the direction of the earth's field turns so as to point its positive face towards the north : the lines of force then penetrate perpendicularly by its negative face. If, at starting, the flux entered by the positive face of the circuit, this movement would at first have the result of reducing the number of lines traversing it; then towards the end of the movement the flux would enter by the negative face.

143. Relative Energy of Two Currents.—To complete the identity of currents and shells, the relative energy of two circuits traversed by currents i, i' must be expressed by

$$W = -ii''L_m, \qquad \S\ 46.$$

The factor L_m has the dimensions of a length and is called the *coefficient of mutual induction* of the two circuits. By definition ($\S$ 46), $L_m i$ is the flux sent by the current i across i', and $L_m i'$ the flux sent by i' across i.

This last deduction from the properties of shells was not *à priori* evident, for it does not necessarily follow, from the fact that two currents act upon a pole, that they act upon each other; e.g., two pieces of soft iron act on a magnet, but taken separately they have no influence upon each other. It was Ampère that discovered the existence of the forces, called *electrodynamic*, exercised between currents.

144. Intrinsic Energy of a Current.—A circuit carrying a current is traversed by the lines of force which it generates, and which form closed curves around the conductor. The figure assumed by these lines in a plane can be shown by the use of iron-filings strewn on a sheet of paper through which the current passes. This figure is analogous to that for a lamellar magnet with the same contour, and magnetized on its opposite faces.

Suppose the circuit be placed in a medium of constant permeability, for example air, and denote by L_i the flux of force passing in the circuit when the current is equal to unity. For a current c the flux will be

$$L_i i = \Phi.$$

Now a current traversed by a flux possesses a reserve of potential energy, the variation of which measures the work performed. In the case we are considering, the flux is dependent on the current ; in order to find the expression for the energy it is necessary to use the method of reasoning employed in regard to the phenomena of electrification or magnetization (§ 25).

When the current varies by di, the flux varies by dΦ, and the potential energy by

$$i\,\mathrm{d}\Phi = L_i i\,\mathrm{d}i.$$

This energy is essentially positive, for the setting up of the current demands an expenditure of energy.

If, then, the current passes from o to c, the energy, which at first is zero, becomes

$$\int_{o}^{i} L_i i\,\mathrm{d}i,$$

or

$$\frac{L_i i^2}{2}.$$

This expression represents the *intrinsic energy of the current.*

The coefficient L_i, whose dimensions like those of the coefficient of mutual induction reduce to a length, is called the *self-inductance*, or *coefficient of self-induction, of the circuit.*

145. Faraday's Rule.—Before passing to the applications of these various formulæ, let us try to find another expression besides Maxwell's (§ 140) for the work done in displacements of a circuit in a field.   Maxwell considers the circuit as a whole, and includes in one simple formula the work done in a deformation or a displacement of the conductors.

It is often useful to analyze separately the action of the various parts of a circuit, and to determine the part which belongs to each one of them in the work accomplisned.

For this purpose let us consider again the expression for the work of an element of current ds (§ 139), which is displaced by ds' in a field of intensity $\mathcal{K} = \dfrac{m}{l^2}$, due to a pole m,

$$dW = i\mathcal{K}ds \sin (l, ds)ds' \cos (df, ds').$$

Now the product

$$\mathcal{K}ds \sin (l, ds)ds' \cos (df, ds'),$$

which represents the product of the field-intensity by the projection of the area described by the given current-element upon a plane normal to the direction of the field, is simply the flux of force swept over by the conductor, for $\mathcal{K}$ is the flux per unit equipotential surface.

It is easy to extend this to a conductor of finite length and to obtain the following rule, first pointed out by Faraday:

The work accomplished by a conductor which is displaced in a field is equal to the product of the current by the flux of force (or number of lines of force) cut by the conductor.

It should be further remembered that the current tends to move towards the left-hand of Ampère's manikin when he faces in the direction of the field.   If the conductor is moved in such a way as to cut no lines of force, the work accomplished is zero.   This is the case when the conductor is displaced parallel to the direction of the field.

APPLICATIONS RELATING TO THE MAGNETIC POTENTIAL OF THE CURRENT.

146. Case of an Indefinite Rectilinear Current.—Let us see whether the application of the idea of potential leads to the expression for the electromagnetic force found by Biot and Savart (§ 136), in the case of an indefinite rectilinear current acting on a neighboring pole.

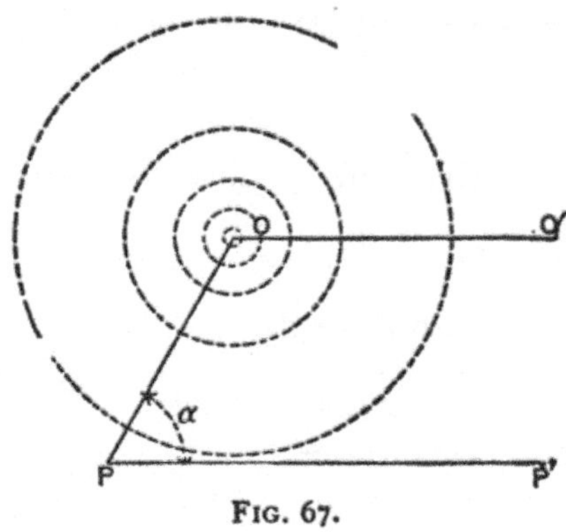

FIG. 67.

Such a current projected on O can be considered as the limit of a plane circuit projected along OO', and extended indefinitely towards the right. The conductors which complete the circuit being infinitely removed from the pole, supposed to be at P, have no action upon it.

Let us now interpret the expression for the potential

$$U = - i(\omega \pm 4\pi n).$$

The solid angle ω, subtended by the circuit at the point P, is obtained by cutting the sphere of unit radius, drawn about P, by a diametral plane OP, which includes all the right lines joining the point P to the conductor projected on O, and by a second plane PP', which likewise includes all the lines drawn from P to the infinitely distant limit of the given imaginary circuit. The portion of the spherical surface thus cut off is a segment which is measured by twice the dihedral angle α between the planes OP, PP'.

We have therefore

$$U = - i(\pm 2\alpha \pm 4\pi n);$$

the sign of 2α being positive or negative according as the current flows upwards or downwards.

The potential has consequently a constant value in the plane OP, which is equipotential.

The intensity of the magnetic field produced by the current is

$$\mathfrak{K} = - \frac{dU}{ds}.$$

In any point of the plane OP the forces of the field are directed perpendicularly to this plane : let us now try to find the value of the intensity in this direction.

Let

$$OP = l ;$$

we have

$$ds = ld(\pi - \alpha) = - ld\alpha,$$

whence

$$\mathfrak{K} = \mp \frac{2id\alpha}{rd\alpha} = \mp \frac{2i}{l}.$$

This expression conforms with Biot and Savart's law. If the current O flows upwards, P tends to approach the shell and the sign of the force is negative; the sign is positive in the contrary case.

The dotted circumferences around the point O show lines of force which correspond to field-intensities decreasing in geometrical progression. The planes passing through the axis of the conductor are normal to the lines of force, and consequently equipotential.

147. Case of a Circular Current.   Tangent-galvanometer.—A circular current of radius R subtends, at a point P on its axis OP, a solid angle ω measured by the spherical segment

$$\dotplus\ 2\pi(1 - \cos\alpha).$$

If, for an observer situated at P, the movement of the current is in the same direction as the hands of a watch, the potential at the point P is

$$U = -i(\omega \pm 4\pi n) = -2\pi i(1 - \cos\alpha \pm 2n).$$

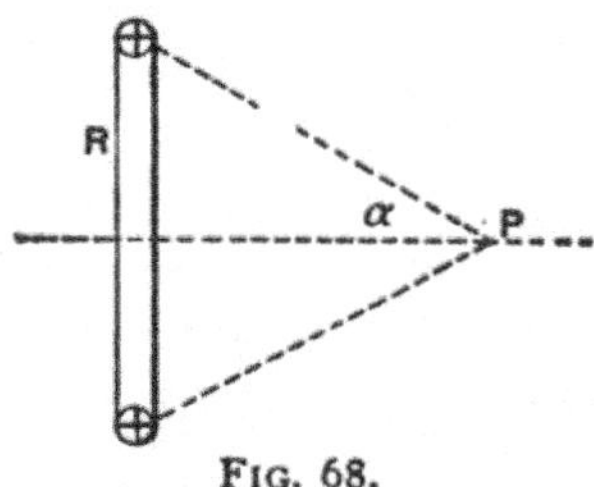

Fig. 68.

The intensity of the field due to the current must be di. rected along the axis OP by reason of symmetry; it is therefore expressed by

$$\mathcal{H} = -\frac{dU}{dr} = +\frac{d}{dr}\,2\pi i\!\left(1 - \frac{r}{\sqrt{r^2 + R^2}} \pm 2n\right)$$

$$= -2\pi i\,\frac{R^2}{(r^2 + R^2)^{\frac{3}{2}}}.$$

That is, on the above hypothesis, P is attracted towards O.

If the point P were at the centre of the circle, the intensity would become

$$\mathcal{H}' = \frac{2\pi i}{R} = \frac{lc}{R^2},$$

l being the length of the current.

If there were *n* circular currents, so near together that their mutual distances were negligible compared to the radius R, the intensity at the centre would be

$$\mathcal{H}'' = \frac{2\pi n i}{R}.$$

Fig. 69 shows the distribution of equipotential lines and lines of force (marked by arrows) in a field due to a circular

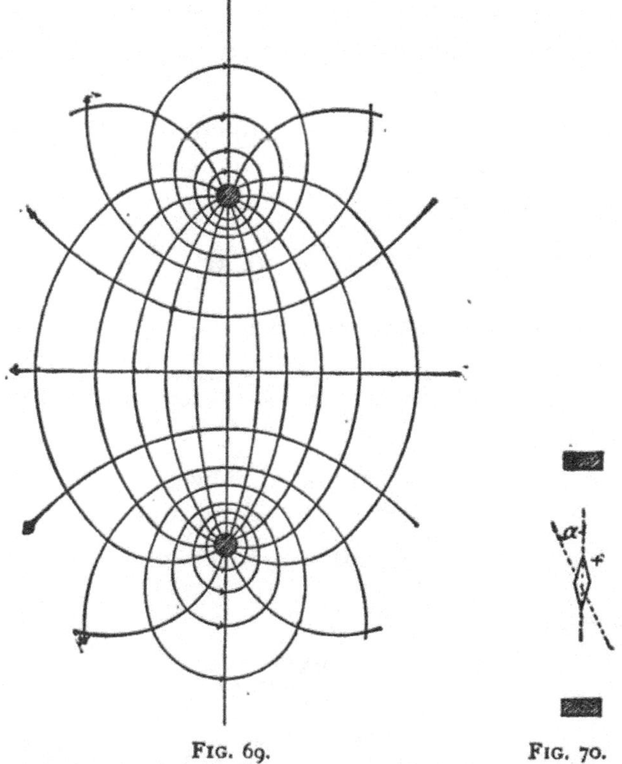

FIG. 69.   FIG. 70.

current. The intensity varies in inverse ratio to the distance between the equipotential lines, and directly as the density of the lines of force.

Suppose a magnet-needle of very small dimensions be

hung by a silk fibre of negligible torsion in the centre of a vertical circular frame, round which are wound n spiral coils of wire, very close together. Suppose, moreover, that the frame be oriented in the plane of the magnetic meridian.

When a current i is sent through the spirals, the needle is acted on by the electromagnetic force, on the one hand, which tends to put it crossways to the current; on the other hand, by the terrestrial magnetism, the action of which opposes this movement.

Under the influence of these contrary actions, the needle takes up a position of equilibrium corresponding to an angle α with the meridian.

Denoting by $\mathfrak{M}$ the magnetic moment of the needle, by $\mathfrak{H}$, the horizontal component of the earth's field, the couple due to the earth is

$$\mathfrak{M}\mathfrak{H} \sin \alpha \quad (\S\ 37).$$

The couple due to the current is

$$\mathfrak{M}\mathfrak{H}'' \cos \alpha = \mathfrak{M}\frac{2\pi n i}{R} \cos \alpha.$$

As these two couples are in equilibrium,

$$\frac{2\pi n i}{R} \cos \alpha = \mathfrak{H} \sin \alpha,$$

whence

$$i = \frac{R\mathfrak{H}}{2\pi n} \tan \alpha.$$

Knowing $\mathfrak{H}$, R, and n, and measuring α by one of the methods shown in § 50, we can deduce from the preceding expression the strength of the current around the frame. This apparatus is called the *tangent-galvanometer*. The spirals of wire are called the galvanometer-coil or multiplier ; the quantity $\dfrac{R}{2\pi n}$ is the *reduction factor* of the galvano-meter.

148. Thomson Galvanometers.—The application of the above simple formula necessitates such a displacement between the coil and the poles of the needle that the tangent-galvanometer is by no means sensitive.

In order to measure weak currents, the multiplier must be wound very close to the needle. In order to determine the best form to give it with a view to economize the wire and diminish the resistance, let us take up again the expression for the action of a circular current on unit pole situated in o, Fig. 71.

$$\mathcal{K} = 2\pi i \,\frac{R^2}{(r^2 + R^2)^{\frac{3}{2}}}.$$

If we put $\mathcal{K} = $ const., and consider r and R as variable, we get the equation of a curve having two symmetrical parts with regard to o, Fig. 71.

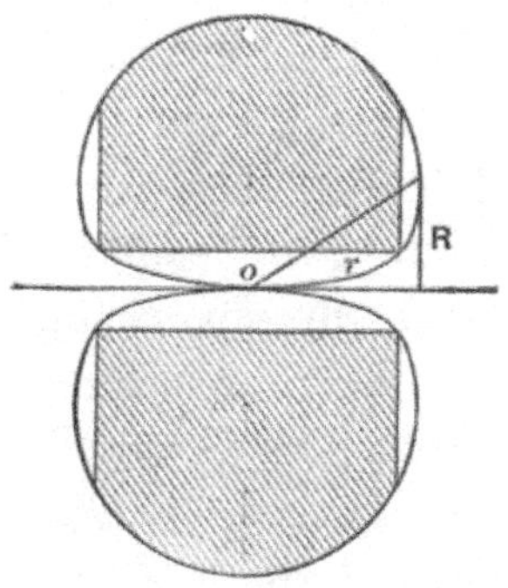

FIG. 71.

The area limited by this curve is the meridian section of a volume of revolution about which the wire can be wound. All the spirals composing such a volume will have an action on the unit pole at least equal to $\mathcal{K}$. All the spirals exterior

to this volume will have a less action.   The cross-hatched
section (Fig. 59) which allows for a cylindrical cavity in
which to place the needle represents, therefore, the rational
form for the coil of a very sensitive galvanometer.   Lord
Kelvin has approached as nearly as possible to this form in
the construction of his galvanometers.

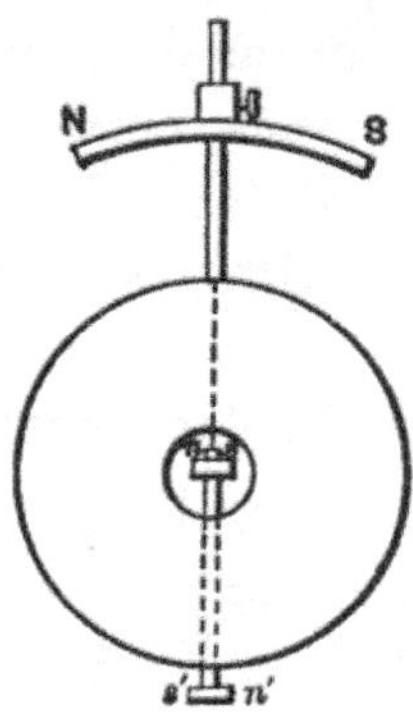

FIG. 72

In order to increase the sensitiveness, it is possible, in addition, to diminish the action of the earth on the needle.   Two
means are employed for this purpose.   The rod which supports the needle and passes through the coil carries a second
needle, *n's'*, oriented in the opposite direction to the first
(Fig. 72).   By choosing two needles of slightly different
magnetic moments, the directive couple due to the earth
can be diminished as much as desired.   Such a system of
of needles is called *astatic*.   The earth's action may likewise
be modified by a *compensating magnet NS*, placed above
the needle.   By changing the position and orientation of
this magnet it is possible to diminish or increase its compensating, or directive, effect at will.   These various methods
are often employed together in Lord Kelvin's galvanometers.

In consequence of the complex form of the coil and the
nearness of the current to the poles of the magnet, the cur-

rent c is not related to the deviation α by a simple formula, as in the case of the tangent-galvanometer.

If i is developed as a function of α, according to Maclaurin's series, it will take the form

$$i = f(\alpha) = f(0) + \frac{\alpha}{1} f'(0) + \frac{\alpha^2}{1.2} f''(0) + \ldots$$

The function must become zero when α is zero, whence it follows that

$$f(0) = 0.$$

Moreover, if the deviation is very small, we can neglect the third and subsequent terms, and take the current as proportional to the deviation

$$i = k\alpha.$$

The coefficient k, which is the reduction factor, is determined by sending a current of known strength through the coils.

The last formula is admissible for deviations of less than $3°$; the readings are made, necessarily, by the method of reflection (§ 50).

149. Shunt.—When the deviation exceeds this limit, it is reduced by an artificial resistance called a *shunt* placed around the galvanometer.

The partial current through galvanometer is then, denoting by g the resistance of the apparatus, and by s that of the shunt (§ 121),

$$ig = i \frac{s}{s+g};$$

whence

$$i = ig \frac{s+g}{s} = mig.$$

The factor m is called the *multiplying power of the shunt.* It is the factor by which the value of the current as found by the galvanometer must be multiplied, in order to obtain the total current.

150. Measurement of an Instantaneous Discharge.— Suppose a quantity of electricity q traverses the coil of a tangent-galvanometer with such rapidity that the needle is not displaced by an appreciable quantity during the discharge. Suppose, too, that the movement of the needle is not damped, so that a double oscillation is obtained having a duration

$$T = 2\pi\sqrt{\frac{\Sigma mr^2}{\mathfrak{H}\mathfrak{M}}}. \qquad (\S\ 37)\ \ .\ \ .\ \ .\ (1)$$

Let us express the fact that the kinetic energy of the needle is equal to the terrestrial couple, and that its momentum represents the impulse that has been given to it.

In expressing these facts it must be remembered that equations relative to movements of translation are applicable to movements of rotation, if the masses be replaced by the moments of inertia, the forces by the couples, and the linear by the angular velocities.

Let ω be the initial angular velocity, and α the maximum deviation of the needle. The terrestrial couple is, for an angle α,

$$\mathfrak{M}\mathfrak{H}\ \sin\ \alpha.$$

The equation of the kinetic energies gives

$$(\Sigma mr^2)\frac{\omega^2}{2} = \int_0^{\alpha_1} \mathfrak{M}\mathfrak{H} \sin\ \alpha\,d\alpha = \mathfrak{M}\mathfrak{H}(1 - \cos\ \alpha_1)$$

$$= 2\mathfrak{M}\mathfrak{H} \sin\ \frac{\alpha_1}{2}.\ .\ \ .\ \ .\ \ .\ \ .\ \ .\ \ .\ (2)$$

Let i be the current of discharge; its action on the needle produces a couple

$$\mathfrak{M}i\,\frac{2\pi n}{L}. \qquad (\S\ 147)$$

The impulse communicated to the needle is

$$\int_0^\tau \mathfrak{M}\,\frac{2\pi n}{L}\,i\,dt = (\Sigma mr^2)\omega, \quad \cdot \quad \cdot \quad \cdot \quad \cdot \quad (3)$$

τ representing the duration of the discharge.

Now

$$\int_0^\tau i\,dt$$

is the quantity of electricity q which was discharged.

Eliminating Σmr^2 and ω between equations (1), (2), and (3), we get

$$q = \frac{T}{\pi}\,\frac{L\mathfrak{K}}{2\pi n}\,\sin\frac{\alpha_1}{2}.$$

In the case of a very small deviation we get simply

$$q = \frac{T}{\pi}\,\frac{L\mathfrak{K}}{2\pi n}\,\frac{\alpha_1}{2} = A\alpha_1.$$

The quantity of electricity is then proportional to the arc of the needle's swing. Under the condition that the deviations are slight, this formula may be extended to any form of mirror-galvanometer whatever.

In order to satisfy the condition at the beginning of this paragraph, galvanometers are chosen for this purpose with a heavy needle having considerable moment of inertia. These needles swing slowly, and enable us to read the limit of the swing exactly.

151. Solenoid. Cylindrical Bobbin.—Under the name of solenoid Ampère defined a series of equal circular currents, placed very close together and normal to a rectilinear or curvilinear axis passing through the centres of gravity of the surfaces which have these currents as their edges.

Denote by s the surface of the circuits, by ϵ their distance apart, and by i the current. Each of them can be replaced by a shell of the same contour, and having a strength i. We may choose the thickness of the shells arbitrarily: we will take it equal to ϵ. Denoting by σ the magnetic density of the faces of the shells, and observing that

$$\epsilon\sigma = i,$$

we have

$$\sigma = \frac{i}{\epsilon} = in_{,}$$

$n_{,}$ representing the number of currents per unit length.

The adjoining faces of neighboring shells counterbalance each other, and these remain, at the extremities of the series, poles whose mass is

$$m = in_{,}s.$$

A bobbin formed on a layer of insulated wire wrapped round a cylindrical core may be considered as a solenoid when the wire is traversed by a current. As a result, however, of the obliquity of the spirals of the bobbin there is an exterior action, which may be obviated by bringing back the ends of the wire along the axis of the bobbin (Fig. 73).

If there are an even number of layers of wire on the core with their spirals inclined in opposite directions, the effects due to their obliquity are rendered null.

Supposing the thickness of the layers to be negligible

compared to their diameter, the resultant poles of the bobbin are given by

$$m = \pm i n_i s,$$

n_i being the number of turns per unit length, and s their mean surface.   The magnetic moment of the solenoid is

$$ml = ins,$$

n representing the total number of turns of wire

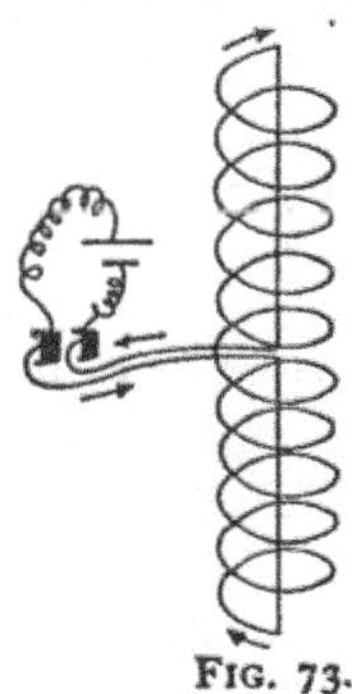

FIG. 73.

Such a solenoid has all the magnetic properties of a uniform cylindrical magnet (§ 42).

The earth's action on the solenoid is shown by hanging it from the ends of the wire (plunged into cups of mercury), which serve as pivots, and at the same time give access to the current from a battery, Fig. 73. By this means it is shown that the positive face of the solenoid turns towards the north.

To determine the intensity of the field due to the solenoid, it need only be remembered that it can be replaced by a uniform cylindrical magnet.

Its action on unit pole situated on its axis, at some external point, is

$$\mathcal{K} = \sigma(2\omega - \omega'), \qquad (\S\ 31)$$

ω and ω' being the solid angles subtended by the bases of the cylinder at the given point.

If the unit pole is situated in the plane of one of the bases, the above equation becomes

$$\mathcal{K} = \sigma(2\pi - \omega').$$

Finally, in order to obtain the expression for the field at a point inside the solenoid, suppose the latter to be cut into two portions by a plane through the given point. The total action is the sum of the actions of the two portions. Now the effect of the first is

$$\sigma(2\pi - \omega),$$

that of the second

$$\sigma(2\pi = \omega').$$

As these actions are added the resultant action is

$$\mathcal{K} = \sigma(2\pi - \omega) + \sigma(2\pi - \omega').$$

In the particular case where the cylinder extends a considerable distance on each side from the point the angles ω and ω' become negligible compared to 2π, and the action is expressed by

$$\mathcal{K} = 4\pi\sigma = 4\pi n_1 i.$$

This expression represents the flux per unit of section taken normally. The total flux is

$$4\pi n_1 is.$$

This flux remains constant in the cylinder up to a certain distance from the ends. When near the ends the intensity of the field decreases, since one of the solid angles ω or ω' assumes an increasing value. The internal flux consequently decreases, and, as in the case of a magnet, lines of force make their exit through the sides of the cylinder. The total flux $4\pi n_{,}is$ is consequently divided into two parts, one emerging from the cylinder by the positive face, the other by the sides. These two sets of lines of force spread out through the surrounding space, and re-enter in the same way by the negative face of the cylinder and the adjoining lateral walls.

We conclude from the above, that inside a cylindrical bobbin of great length a uniform magnetic field is produced, directed parallel to the axis of the cylinder from the S to the N face; the intensity of the field is measured by 4π, multiplied by the product of the current into the number of spirals turns in unit length.

Such a bobbin consequently furnishes a practical means of obtaining a uniform field whose intensity is only limited by the heating of the wire by the current.

The identity of the external effects of solenoids and magnets has led to the conclusion that there is an analogy in their internal effects, which latter cannot be directly determined in the case of magnets. It is known, indeed, that if a cavity be made in a magnet, its walls form poles whose effect is added to that of the end-poles. Care must be taken, however, not to confuse the internal action of a solenoid with that of a tubular magnet, for in such a magnet the lines of force have the same direction inside and outside, their return taking place in the thickness of the tube.

To sum up, it is admitted that magnets, like solenoids, gives a constant and continuous total flux, which makes its exit by the N end, and returns to the point of departure by entering through the S end. We shall see that this consideration has led to a comparison between magnetic and electric fluxes, and to treating the former by relations analogous to the latter.

152. Electrodynamometer.—The magnetic properties of solenoids have been carefully verified by Weber, who by means of them has repeated Gauss' experiments on magnetism (§ 48). To give the desired mobility to one of the solenoids, it is hung by two slender wires, which at the same time serve for the entrance and exit of the current.

The solenoid may be supported by two wires placed side by side: in this case the torsion-couple which balances the mutual action of the two solenoids is proportional to the sine of the angle of torsion. If the suspension wires are one above and one beneath the solenoid in the same right line, the moment of torsion is simply proportional to the angle of torsion, and one of the wires supports the whole weight of the movable solenoid.

The mutual action of the two solenoids is proportional to their magnetic moments, and, consequently, to the product of the currents which traverse them. If the same current passes through both, the couple is proportional to its square.

Weber has utilized these properties in the construction of the *electrodynamometer*. In this apparatus, which serves to measure the current-strength, the movable solenoid is hung at the centre of the fixed one and at right angles with it.

The current to be measured passes successively through both solenoids, whose axes are then urged to assume a parallel position. The turning couple, balanced by the torsion of the suspension wires, is proportional to the square of the current. This property allows the apparatus to be applied to the measurement of currents whose flux varies periodically in direction, since the mutual action of the two solenoids preserves the same sign, whatever be the sign of the current.

The action of terrestrial magnetism on the movable

solenoid must be taken into account. The solenoid can, however, be reduced to a very small number of spirals, so as to render this effect of but little account.

A more exact method consists in first adjusting the movable solenoid so that its axis is in the magnetic meridian, with the positive end turned towards the north. When the solenoid is deflected under the influence of a constant current, it is brought back to its initial position by turning the upper part of the suspension wire. The angle of torsion enables us to measure the deflecting current, the terrestrial couple being zero.

The angle of torsion β is proportional to the square of the current through the two solenoids, or

$$\beta = ki^2.$$

153. **Case of a Ring-shaped Bobbin or Solenoid.**—Suppose a layer of wire wound so as to form a ring of rectangular section (Fig. 74), each spiral being situated in a meridian section. The magnetic effect of such a bobbin, traversed

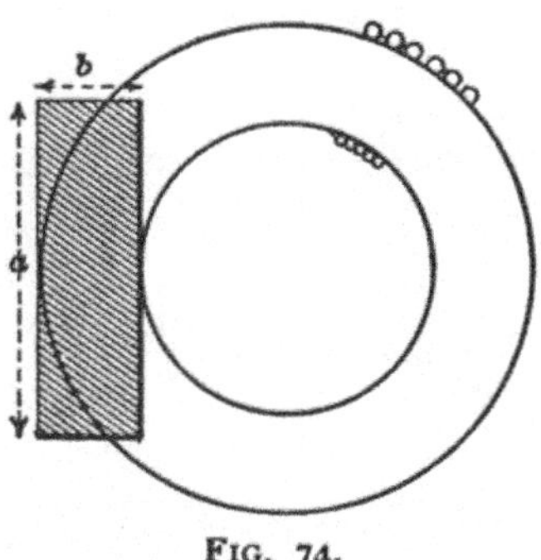

FIG. 74.

by a current, is zero in all external points. The internal flux of force created by the system is composed of lines of force concentric with the ring. This system is equivalent to a magnet formed of magnetic filaments closed on themselves.

The field is variable in a meridian section of the ring. It is easy to see, indeed, that the number of turns of wire per unit length is smaller on the outside than on the inside of the ring; it follows that the field within the ring is more intense at the inner periphery than at the outer.

In order to determine the intensity of the interior field, denote by n' the number of spirals comprised between two meridian planes whose angular opening equals one radian. At a distance r from the axis the space between two adjoining spirals is $\epsilon = \dfrac{r}{n'}$. The intensity of magnetization of the equivalent magnetic solenoid is

$$\mathfrak{I} = \frac{i}{\epsilon} = \frac{n'i}{r};$$

corresponding to this intensity, per unit surface, is a flux equal to

$$4\pi\mathfrak{I} = 4\pi\frac{n'i}{r}.$$

Across an element of section ds, at a distance r from the axis, the flux is

$$\mathrm{d}\Phi = 4\pi n'i\frac{\mathrm{d}s}{r}.$$

The total flux across a section of the ring whose height is a, thickness b, and interior radius R, is

$$\Phi = 4\pi n'i \int_{R}^{R+b} \frac{\mathrm{d}s}{r} = 4\pi n'i \int_{R}^{R+b} \frac{a\,\mathrm{d}r}{r} = 4\pi n'ia \, \log. \frac{R+b}{R}.$$

Calling n the total number of spirals and n_1 the number per unit length along the inside face of the ring, we have

$$n' = \frac{n}{2\pi} = n_1 R,$$

whence

$$\Phi = 2nia \log_e \frac{R+b}{R} = 4\pi n_1 iRa \log_e \frac{R+b}{R}.$$

The mean intensity of the field inside the ring is

$$\mathcal{H}_m = \frac{\Phi}{ab} = \frac{4\pi n_1 iR}{b} \log_e \left(1 + \frac{b}{R}\right).$$

It is easily proven that this expression becomes $4\pi n_1 i$ for a ring of very great interior diameter.

In fact, developing the logarithm in series, we get

$$\mathcal{H}_m = 4\pi n_1 i \frac{R}{b} \left(\frac{b}{R} - \frac{b^2}{2R^2} + \frac{b^3}{3R^3} + \cdots\right)$$

$$= 4\pi n_1 i \left(1 - \frac{b}{2R} + \frac{b^2}{3R^2} + \cdots\right).$$

If b is negligible compared to R, we get simply

$$\mathcal{H} = 4\pi n_1 i.$$

In the case of a ring of circular section, where the radius of the circular axis of the ring is R, and that of a meridian section a, we get

$$\Phi = 4\pi n_1 iR \int_{R-a}^{R+a} \frac{ds}{r} = 8\pi^2 n_1 iR(R - \sqrt{R^2 - a^2})$$

Hence a mean internal intensity equal to

$$\mathfrak{K}_m = \frac{8\pi n_1 iR}{a^2} \left(R - \sqrt{R^2 - a^2} \right).$$

Developing the expression under the radical sign into a series and neglecting the ratio $\frac{a}{R}$ in the case of a ring of great diameter, we would have simply

$$\mathfrak{K}_m = 4\pi n_1 i.$$

ELECTROMAGNETIC ROTATIONS AND DISPLACEMENTS.

154. An invariable system of electric currents is not capable of producing the rotation of a magnet, for if the magnet returns to its initial position after having traversed a current c, the work accomplished by its poles of masses $+ m$ and $- m$ is

$$+ 4\pi mi - 4\pi mi = 0. \qquad (\S\ 141)$$

If the magnet does not traverse the current, the work is equally zero, so that in no case can the resistance due to friction be overcome.

But rotation can be produced by various artifices, such as rendering only one part of the electric or magnetic system movable, changing the direction of the current, etc.

155. **Rotation of a Current by a Magnet.**—A vertical magnet serves as pivot to a balanced conductor whose lower end plunges into a circular trough of mercury which extends round the middle of the magnet. A voltaic element is placed in the circuit between the trough and the magnet so as to furnish a steady current. According to

Faraday's rule (§ 145), the movable conductor tends to move so as to cut the lines of force due to the magnet.

In the course of one revolution the conductor cuts all the

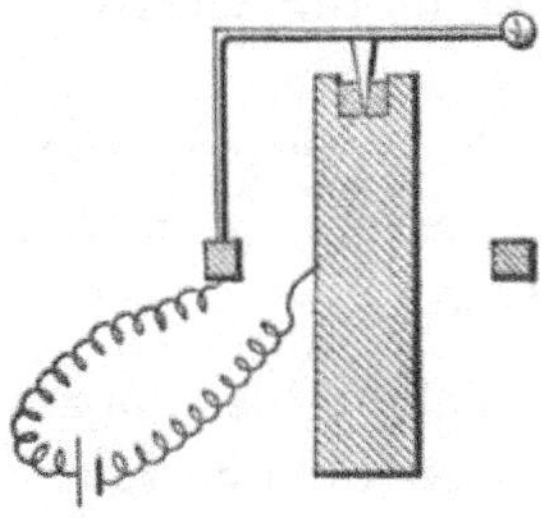

FIG. 75.

lines of force emanating from one of the poles m, or $4\pi m$ lines, according to Gauss' theorem. The work accomplished is therefore (§ 145)

$$w = i \times 4\pi m.$$

The couple acting on the conductor is

$$\frac{w}{2\pi} = 2mi.$$

156. This experiment can be repeated by the aid of an electric discharge in the form of a brush-discharge, obtained

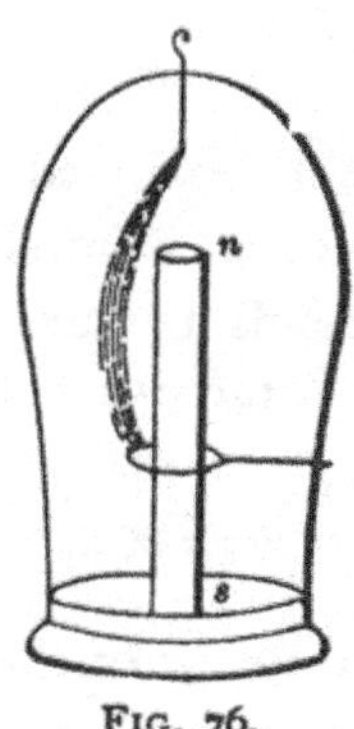

FIG. 76.

by maintaining two conductors, a, b, at a considerable differ-

ence of potential inside a vessel from which the air is par-
tially exhausted.   If the conductor *a* passes round the middle of the magnet *ns* the luminous band is seen to turn about
the magnet by reason of the electromagnetic force which
impels it.

A liquid traversed by a current and situated in a magnetic field will likewise assume a movement normal to the
lines of force of the field.

157. Barlow's Wheel.—This apparatus gives another
example of the rotation of a current under the influence
of a magnet.   It consists of a copper wheel whose metallic

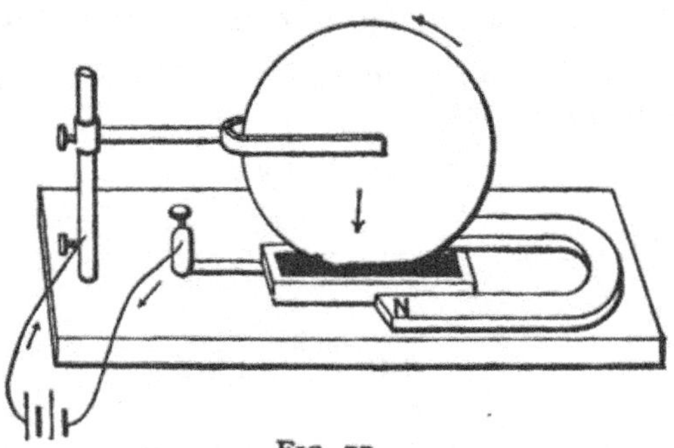

FIG. 77.

axis conducts the current to the wheel, from which it passes
into a trough of mercury, into which the lower edge of the
wheel dips.

The part of the wheel traversed by the current is embraced between the poles of a horseshoe magnet, *N*.   The
electromagnetic force tends to turn the wheel in the direction of the arrow.   By Faraday's rule (§ 145), the work per
revolution is

$$w = \mathcal{H}si,$$

$\mathcal{H}$ denoting the mean intensity of the field cut by the wheel
s the surface described by one of its radii—that is, the surface of the wheel itself, and *c* the current.

158. Rotation produced by Reversing a Current.—The various preceding examples suppose no variation in the direction of the current ; in the following case this reversal is the cause of rotation. Suppose a bobbin movable round an axis perpendicular to its own, in a field which, for simplicity, we will suppose uniform.

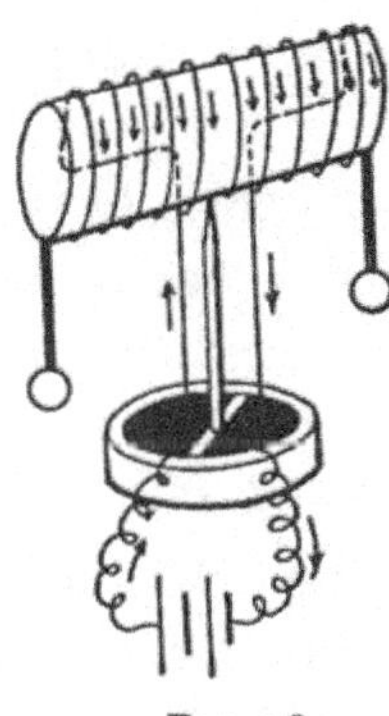

FIG. 78.

The ends of the insulated conductor which is wrapped on the bobbin dip into a trough divided by a partition parallel to the field into two compartments filled with mercury. If a current is passed through the mercury into the bobbin, the latter tends to orient itself so as to embrace the greatest possible flux by its negative face (§ 142). But at the moment when this position is reached, the moving contacts cross the partition in consequence of the acquired velocity; the current is then reversed in the bobbin, the electromagnetic couple preserves the same sign, and the movement continues as long as the current.

Denoting by s the mean surface of the spirals, by n their number, and by $\mathcal{K}$ the field-intensity, the work accomplished in one revolution is (§ 142)

$$i \times 4\mathcal{K}ns.$$

159. Mutual Action of Currents.—The form of the field produced by a current being given, we need only apply Ampère's rule (§135) to predict the reaction of two currents. It is readily seen that two parallel currents attract each other when they have the same direction, and repel each other if they have opposite directions.

In consequence of this the spirals of a solenoid tend to approach each other when they are the seat of an electric flux.

Two angular currents attract or repel each other according as they are directed in the same or opposite directions when referred to the vertex of the angle.

Take the particular case of two parallel currents i, i', the one of finite length l, the other indefinite; let r be their distance apart. The field due to the indefinite current c, in every point of the other conductor, has an intensity

$$\mathcal{H} = \frac{2i}{r} \qquad (\S\ 146)$$

The electromagnetic force directed along r will give for a displacement dr of the current i' an element of work

$$dw = i'\mathcal{H}l\,dr; \qquad (\S\ 145)$$

consequently the force acting between the currents is

$$f = \frac{dw}{dr} = i\mathcal{H}l = 2\frac{ii'}{r}\,l.$$

160. Reaction produced in a Circuit traversed by a Current.—The lines of force generated by a current across its own circuit naturally penetrate by the negative face, and their electromagnetic action upon the current consequently tends to increase the surface limited by the conductors.

If, then, a circuit of irregular form is traversed by a current, the conductors tend to be made to assume a circular shape, which corresponds to the maximum surface. To render this effect evident, however, the current must be sufficiently strong to heat the wires enough to soften them and enable them to obey the electromagnetic forces.

FIG. 79.

The reaction of a current upon itself can be shown in another way. A circuit (Fig. 79) is completed by a curved movable conductor floating on mercury contained in two long troughs. When a current is set up, the movable conductor is displaced towards the right, so as to increase the surface of the circuit.

161. Explanation of Electromagnetic Displacements based on the Properties of Lines of Force.—Faraday accounted for magnetic or electromagnetic actions by attributing certain properties to the magnetic lines of force.* Instead of admitting, as did the physicists who preceded him, that currents and magnets act at a distance without intermediary, he was convinced that their reactions are produced by means of a medium in a state of tension or movement such that the currents and magnets thus undergo the observed actions.

As we have seen in § 47, Faraday imagined the medium as tense along the lines of force, for which he substituted,

---

* Faraday, *Experimental Researches.*

in thought, elastic threads *having a tendency to shorten, while choosing the path that is most permeable magnetically.*

The curved form of the lines of force is explained in this hypothesis by the *mutual repulsion of neighboring lines having the same direction.* On the other hand, *if the neighboring lines have opposite directions, they attract each other and tend to unite.*

These practical rules suffice to predict all electromagnetic actions. Given a system of magnets and currents, it is always possible to represent mentally the distribution of the lines of force due to each of the elements of the system. On then combining these lines according to the directions given above, we obtain the form of the resultant field, to which we need only apply the first of the above rules in order to determine the direction of the displacement which will occur.

But it is easier to find the resultant field directly by using the method of magnetic figures (§ 47). The iron-filings show the distribution of lines of force, which take the form of closed curves around currents and of interrupted lines in magnets. By their tendency to shorten, the former are urged to assume a circular shape as having a minimum perimeter, and the latter a rectilinear form. By means of these magnetic figures the movements of the system can be clearly foreseen.

As an example, let us consider a small straight magnet capable of turning between the limbs of a horseshoe magnet (Fig. 80). The iron-filings figure shows that the lines of force of the horseshoe magnet are bent, and that a portion of them enter the small magnet by its *s* end, leaving by its *n* end. Their tendency to shorten causes the small magnet to place itself in the direction of the line joining the poles of the other magnet.

The same thing happens if we replace the straight mag-

net by a circular current (Fig. 81).   The lines of force from
the horseshoe magnet will traverse the current through its

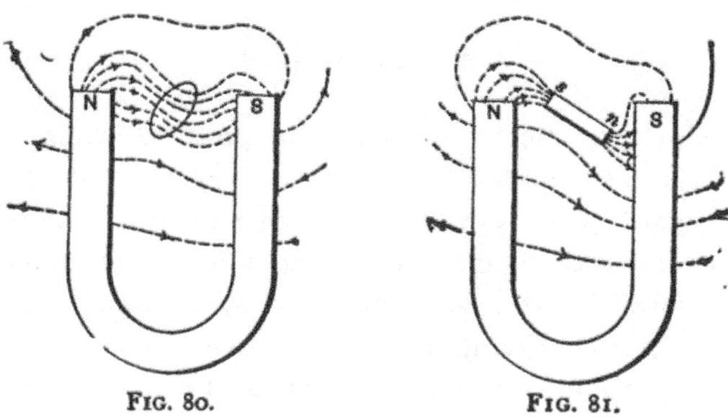

Fig. 80.Fig. 81.

negative face and force it to assume a position normal to
the line between the poles.

In the case of Barlow's wheel (§ 157), the lines of force
from the magnet are shifted by those of the current traversing the wheel.   The result is a field whose lines embrace
the movable disc and pull it in one direction or the other,
according to the direction of the current.

So, too, two parallel currents in the same direction combine their fields in such a way as to furnish lines which embrace both conductors, and oblige them to approach each
other (Fig. 82).

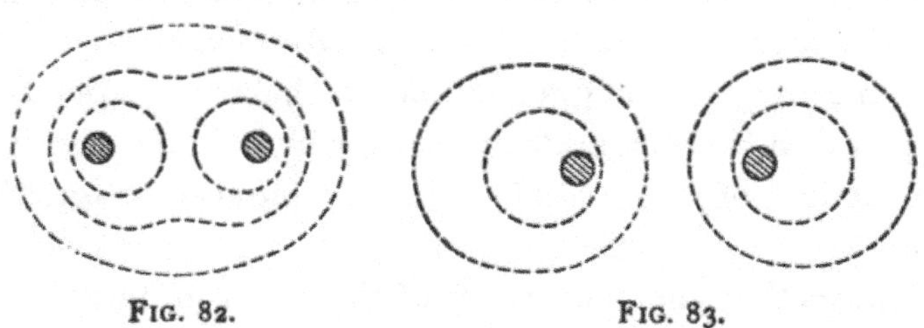

Fig. 82.Fig. 83.

If the currents are in opposite directions, the circular lines
of force surrounding each conductor assume an oblong

shape : to enable them to return to their circular form, it is
necessary for the currents to move apart.

ELECTROMAGNETS.

162. Since an electric current can produce intense magnetic fields, as appears from the investigation of straight
and annular solenoids (§ 151), it enables us to obtain permanent or temporary magnets of considerable magnetic power.
Thus a core of soft iron, surrounded by a bobbin of insulated wire, forms a temporary magnet when a current traverses the bobbin. A steel core after the passage of the
current preserves a permanent magnetization in proportion
to the degree of hardness of the metal.

The system composed of this core and bobbin is called an
electromagnet. The poles of the electromagnet are of the same
sign as the corresponding ones of the magnetizing bobbin.
If two *straight electromagnets*, parallel and placed in oppo-

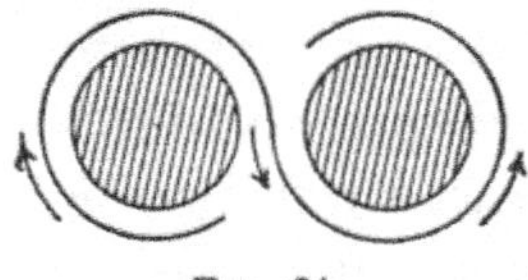

Fig. 84.

site directions, be connected by a cross-piece of soft iron
called the *yoke*, we obtain a *horseshoe electromagnet.*

The intensity of magnetization of the core depends at
each point on the magnetizing force, which is the resultant
of the action of the bobbin and the induced magnetism
(§ 52).

It follows that the magnetism of the core can only be calculated in certain simple cases. If, for example, the bobbin
is straight and indefinitely long, as well as the core, the
field due to the bobbin is constant in the medial region, and

the effect of the induced poles towards the ends of the core is zero in this region.

If we denote by n_1 the number of turns of wire per unit length along the axis of the bobbin, by i the current, the field due to the current is expressed by (§ 151)

$$\mathcal{H} = 4\pi n_1 i.$$

The magnetic induction in the core is (§ 60)

$$\mathcal{B} = \mu\mathcal{H} = (1 + 4\pi\kappa)\mathcal{H}.$$

Let S be the section of the bobbin, S' that of the core; the total flux across the bobbin will be expressed by

$$\Phi = 4\pi n_1 i(S + 4\pi\kappa S').$$

In the case of an annular core (§ 153) we have an identical expression, when the thickness of the ring is small relatively to its diameter.

Towards the ends of a straight electromagnet the flux across its different normal sections is not constant; the intensity of magnetization of the core is feebler there than in the middle on account of the demagnetizing action of the poles. This can be shown by the figures formed by iron-filings, which show lines of force emerging laterally from the core, beginning at a certain distance from the extremities.

163. Energy expended in Electromagnets. General Definition of the Coefficient of Self-induction of a Circuit.—We have seen that the energy of a current due to its field is expressed by

$$\frac{L_1 i^2}{2}, \qquad (\text{§ 144})$$

L_1 denoting the flux traversing the circuit for a current equal to unity, on condition that the permeability of the medium is constant.

In the case of a very long bobbin, the flux across one of its spirals is $4\pi n_1 s$. If there are n spirals and if we neglect the irregularity of the field towards the extremities, we have

$$L_s = 4\pi n_1 ns.$$

The intrinsic energy is

$$2\pi n_1 nsi^2.$$

When an iron core is introduced, of section s, the flux corresponding to a current i becomes approximately

$$4\pi n_1 n(s + 4\pi\kappa s')i,$$

and the energy

$$2\pi n_1 n(s + 4\pi\kappa s')i^2.$$

The difference

$$2\pi n_1 n(4\pi\kappa s')i^2 = \tfrac{1}{2}\kappa\mathcal{H}^2 \times s'l'$$

corresponds to the work of magnetization of the core.

We can also designate by the name of coefficient of self-induction of the electromagnet the relation $L_s' = 4\pi n_1 n$ $(s + 4\pi\kappa s')$ of the flux to the current; but it will be observed that this relation is no longer constant, as in the case of a simple bobbin: it contains the parimeter κ, which is variable with the current.

In the case where a section of the bobbin can be considered as equal to that of the core, we have

$$L'_s = 4\pi n_1 n\mu s.$$

The coefficient of self-induction is then proportional to the permeability of the core. The coefficient of self-induction of a coreless bobbin is invariable, because the permeability of air may be considered as constant and equal to unity (§ 60).

In a general way we see that the coefficient of self-induction of a circuit is the ratio of the flux of magnetic force across it to the current; we cannot define the coefficient as equal to the magnetic flux for unit current unless the circuit is remote from any object made of iron, nickel, or cobalt.

From the moment when the core acquires its normal magnetization, which depends on the current and its previous magnetic conditions (§ 57), the effect of the current is limited to maintaining the magnetization, *i. e.*, the molecular orientation of the core, and to heating the wire in the bobbin.

It is to be noted that a solenoid of given volume can be composed of a small number of turns of coarse wire, or a large number of fine wire. If we neglect the space taken up by the insulation on the wire and the interstices between the coils, we find that the electromagnetic action and the heating are constant as long as the current remains proportional to the cross-section of the wire. In fact, the magnetizing force is proportional to the product of the number of turns by the current; now the first of these quantities is in inverse ratio to the cross-section of the wire, while the second is proportional to it. On the other hand, the heat produced in one second is measured by the product of the square of the current into the resistance of the solenoid, but with a constant volume the total resistance is in inverse ratio to the square of the cross-section of the wire; consequently the product remains constant.

We arrive at the same result by considering the *density of the current* in the solenoid, that is to say, the current per unit of section of the conductor. The magnetizing force and the heating are respectively proportional to the first and second powers of the density: these quantities are therefore constant as long as the density is invariable.

As the heating effect increases much more rapidly than the magnetic effect, it will be evident that it is the heating which limits the results that can be obtained with a solenoid.

164. Magnetic Circuit. Magnetomotive Force. Magnetic Resistance or Reluctance.

—A knowledge of the magnetic induction across electromagnets is very important as regards their application in the construction of machines.

To determine the flux of magnetic force, it must be remembered that the lines of force generated by a current are continuous curves closed upon themselves in a homogeneous or heterogeneous medium (§ 151). In the case of an annular core, completely surrounded by a solenoid, the lines are concentrated in the core (§ 153). In the case of a straight electromagnet the flux of force traversing the solenoid makes its exit from the positive region, and returns by the negative region after diffusing itself through the surrounding space.

The law of continuity, which is obeyed by the magnetic flux as well as by the electric flux or current, leads us to consider whether the two categories of phenomena are not governed by the same laws. We have seen that different substances conduct the lines of magnetic force differently, and we have given the name *permeability* to the coefficient which characterizes a body in this respect. Hopkinson and Kapp have compared this coefficient to the coefficient of conductivity for the electric flux. They have thus been led to apply to the *magnetic circuit*, traversed by a greater or less flux according to its permeability, similar relations to those which govern the current in an electric circuit.

Take the simple case of a homogeneous circuit formed of an iron ring completely surrounded by a solenoid traversed by a current. The magnetic induction is

$$\mathfrak{B} = \mu \mathcal{H}, \text{ where } \mathcal{H} = 4\pi n_{\text{\tiny 1}} i. \qquad (\S\ 153)$$

The total flux across the core, of section s, is

$$\Phi = \mathfrak{B}s = 4\pi n_{\text{\tiny 1}} i \mu s.$$

Calling l the length of the circular axis of the ring, n the total number of turns in the solenoid,

$$n_{\text{\tiny 1}} = \frac{n}{l},$$

whence

$$\Phi = \frac{4\pi n i}{\dfrac{l}{\mu s}}. \qquad \ldots \ldots \quad (1)$$

This form is analogous to Ohm's law; the magnetic flux is proportional to the expression $4\pi n i$, which by similitude is called *magnetomotive force** and inversely proportional to $\dfrac{l}{\mu s}$, which is given the name of *magnetic resistance* of the circuit, on account of its likeness to electric resistance ($\S$ 115).

Heaviside, in consideration of the fact that there is in magnetism nothing analogous to the Joule effect, which is work accomplished by the electric current to overcome the electric resistance, has substituted for the expression "magnetic resistance" that of *reluctance*.

We can get equation (1) by a direct proceeding. We have seen ($\S$ 141) that the work performed by a unit pole

---

* Bosanquet, *Philosophical Magazine*, 1883, Vol. XV, p. 205.

in moving across a circuit is equal to $4\pi i$ times the number
of revolutions made, $4\pi i$ representing the difference of magnetic potential between the two faces of the circuit.　Now
if the unit pole moves along the internal axis of the ring, it
traverses the current n times in one revolution.　The work
accomplished, $4\pi n i$ is also expressed by the product of the
mean intensity $\mathcal{H}$ into the path l traversed by the unit pole,
whence

$$4\pi n i = \mathcal{H}l\,;$$

but

$$\mathcal{H} = \frac{\Phi}{\mu s},$$

whence

$$\Phi = \frac{4\pi n i}{\dfrac{l}{\mu s}}.$$

The magnetomotive force $4\pi n i$ is therefore measured by
the sum of the differences of magnetic potential produced
in the solenoid.

If the core were composed of various segments having
lengths l, l', l''. sections s, s', s'', and permeabilities μ, μ', μ'',
we would have

$$4\pi n i = \Phi\left(\frac{l}{\mu s} + \frac{l'}{\mu' s'} + \frac{l''}{\mu'' s''}\right).$$

There is, however, an essential distinction to be made between the electric and magnetic circuits.　The resistance of
the former is independent of the current, while that of the
latter is a function of the permeability, which depends not
only on the actual flux, but also on the preceding fluxes
(§ 57).　On the other hand, the flux and the quantity of
magnetism are not connected by a law similar to that which
connects the quantity of electricity to the current and the

time. Finally, in consequence of the residual magnetism, a magnetic flux can exist in a circuit without any magnetomotive force, or even a flux in a direction opposed to the magnetomotive force. We must therefore be careful not to conclude an analogy of fact from an analogy of form, and only look on the extension of Ohm's law to the magnetic circuit as an artifice to facilitate the investigation of the question as well as the calculations connected with it.

Following this order of ideas, we can push the comparison still further and treat cases of complex magnetic circuits by Kirchhoff's laws.

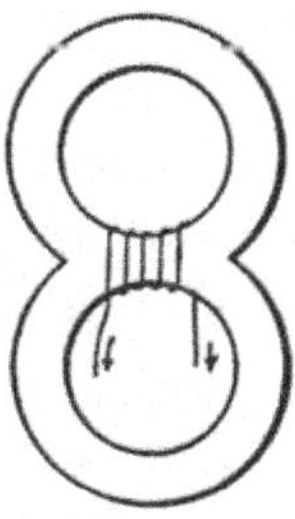

FIG. 85.

Let us take the case of a magnetic circuit, placed in a medium supposed to be impermeable to the lines of force, and divided into two portions. The flux Φ generated by a solenoid of n turns traversed by a current i, divides itself into two derived fluxes, Φ' and Φ''.

We shall have

$$\Phi = \Phi' + \Phi''.$$

Let l be the length of the common branch, s its section, and μ its permeability.

Let l', s', μ' ; l'', s'', μ'' be the corresponding elements for the derived branches.

Kirchhoff's second law will give the equations

$$4\pi ni = \Phi \frac{l}{\mu s} + \Phi' \frac{l'}{\mu' s'},$$

$$4\pi ni = \Phi \frac{l}{\mu s} + \Phi'' \frac{l''}{\mu'' s''}.$$

The case that we have just considered is imaginary, for there is no medium in existence which is impermeable to lines of force. In the phenomena of the electric current we practically take air at the ordinary pressure as a perfect insulator for the electric flux; but the same does not hold for magnetic phenomena. If the permeability of air and other gases is negligible compared with that of iron when traversed by a mean magnetic induction, it becomes decidedly comparable with it for intense magnetic fluxes. It follows that in the case of Fig. 85, for example, a part of the flux generated by the solenoid will go off into the surrounding medium. The ratio of the flux into the air to the flux traversing the iron rings will increase with the current in the solenoid in consequence of the gradual weakening of the permeability of iron and the constancy of that of air.

The case is not without analogy with that of an electric circuit placed in the midst of a liquid having a certain relative conductivity. A part of the current furnished by the electric generator will pass through the liquid without following the line of the metallic circuit. We shall see, when studying dynamos, how the derived flow of the magnetic flux in the surrounding medium can be experimentally estimated.

165. Forms and Construction of Electromagnets.— When it is desired to utilize the portative power of an electromagnet, the horseshoe form is used, consisting of two straight *cores* wound with wire and connected at the base by an iron *yoke.* The free poles attract the *armature* (or *keeper*), which, with the *air-gap,* completes the magnetic circuit.

The attractive force is proportional to the cross-section and the square of the magnetic induction. We have seen (§ 56) that the portative power of an electromagnet per unit section is $2\pi\mathfrak{J}^2 + \mathfrak{IC}\mathfrak{J}$, which expression becomes $\dfrac{\mathfrak{B}^2}{8\pi}$ if we add the very small term $\dfrac{\mathfrak{IC}^2}{8\pi}$. It is therefore advantageous to make electromagnets of short and massive pieces.

When we wish to calculate an electromagnet to obtain a given portative power

$$F = \frac{\mathfrak{B}^2 s}{8\pi},$$

we take a magnetic induction $\mathfrak{B}$, which can be obtained without too great expenditure of electric energy. The magnetism-curves (§ 57) show that beyond $\mathfrak{B} = 16000$ the increase in induction is very feeble relatively to the expenditure: consequently we choose a section s about

$$\frac{8\pi F}{16000^2}.$$

Next we seek the dimensions of the bobbins capable of producing the flux $\mathfrak{B}s$; we take a given length, l, for the cores, and, after having joined them by a yoke and a keeper of section s, we calculate the magnetomotive force $4\pi ni$ that the coils must develop, by the formula

$$4\pi ni = \mathfrak{B}s \times \frac{l}{\mu s}.$$

An example of such a calculation will be seen further on. Instead of using the horseshoe form in order to obtain a closed circuit, the helix of a straight electromagnet may be enclosed by an iron sheath joined to the core at one end by the yoke, and at the other by the keeper, Fig. 86. This

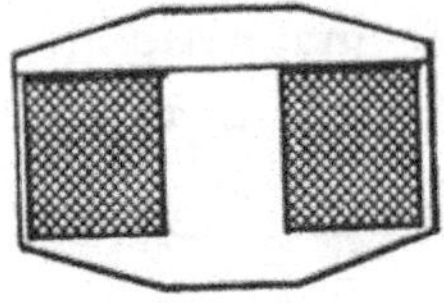

FIG. 86.

form is very compact, but it does not lend itself readily to the employment of strong currents, for the radiation of the heat produced in the helix by the Joule effect does not take place so well as in the other form.

When the keeper is moved away from the poles, the attractive force diminishes very rapidly. In fact the attraction is inversely proportional to the square of the distance between the poles (of the keeper and magnet) which are opposite each other, and the poles themselves decrease with the magnetic flux, which grows very suddenly less on account of the feeble permeability of the air. It follows that at quite a short distance the attraction exercised by the magnet-poles on the keeper becomes negligible; we have, consequently, to limit the amount of play of this movable part.

In order to diminish the weakening of the flux, caused by the large reluctance of the air-gap, we can increase the magnetomotive force by increasing the number of turns in the helix. This leads to the lengthening of the cores; this lengthening produces in itself an increase in the magnetic resistance, but in view of the high value of the permeability of iron compared with that of air, the total resistance of the circuit is not notably increased. This means is made use of in telegraphic electromagnets.

In many applications it is necessary to increase the play

of the keeper; this is managed very simply by giving it, instead of a movement perpendicular to the line of the poles, an oblique movement, which augments its play without altering the work done in displacement. One solution consists in furnishing the keeper with conical projections which penetrate into conical cavities cut in the cores.

If it is desired to obtain considerable displacements, with a small variation in the attractive force, we must make use of the suction-effect of a long solenoid upon a cylindrical core placed at the open end of the solenoid. The core tends to place itself in such a position that the flux traversing it is a maximum (§§ 65, 151).

Before analyzing the effect of a long bobbin or solenoid, placed vertically, upon a core acting as a plunger, it should be recalled that the field is sensibly uniform in the middle region of the solenoid, and that its intensity decreases rapidly towards the ends (§ 151). A pole of invariable strength is therefore urged through the solenoid by a force which increases from the point of entry, becomes constant in the middle, and decreases towards the lower end. The force is in each point equal to the intensity of the field multiplied by the strength of the given pole.

When a soft iron core is presented at the opening of a solenoid, the phenomenon is more complex. The attractive effort depends on the magnetization induced in the core, which itself varies with the intensity of the field. To simplify the matter, let us consider a cylindrical core whose length is very great compared to that of the solenoid, so that we have only to consider the action exercised upon the pole induced in its lower end. The strength of this pole increases as the core descends, for the total flux produced by the solenoid increases in consequence of the gradual diminution of the resistance of the magnetic circuit due to the insertion of the iron core. When, however, its lower end reaches the bottom of the solenoid the force tends to decrease, for the induced pole reaches a region of the field where the intensity is diminishing.

If the length of the core equals that of the solenoid, we must take into consideration the antagonistic action exercised on the pole induced in the upper end of the core. Experiment shows that in this case the resultant effort is maximum when this latter reaches the middle of the solenoid; it then decreases and becomes zero when the core is in a symmetrical position with regard to the solenoid, this being the position in which the flux produced by the current and traversing the iron is a maximum.

In order to regulate the attractive effort exercised on the core, when this latter is subject to considerable displacements, it is given the form of a very long cone with the apex pointing downwards. In this case the total flux increases as the core enters into the solenoid, even when the apex of the cone has passed the lower opening.

Solenoids develop a very much lower attractive effort on their cores than that exercised by electromagnets of horseshoe shape on their keepers, for in the latter case the magnetic circuit has much less reluctance.

To increase the suction-effect we can put an iron sheathing on the solenoid, which presents a very permeable path for the lines of force. A solenoid armored in this way gives a very intense, uniform field in its interior, but its external effects are negligible: the core is not attracted, in this case, until it is introduced into one of the openings in the sheathing.

In order to reduce as much as possible the perimeter of the wire in an electromagnet, the core must have a circular section; sometimes, however, a square or rectangular section is chosen to gain compactness. Such a shape increases not only the length and consequently the resistance of the magnetizing wire, but also the losses of flux which take place between the lateral faces of the adjoining cores, particularly in horseshoe electromagnets, by bringing the cores closer together and increasing their surface.

As far as possible sharp edges on the polar surfaces must be avoided, since the lines of force have a tendency to crowd together at the edges and escape by them into the surrounding air. This effect is not without analogy with the effect of points in the case of electrified bodies.

The wire used in electromagnets is generally of the purest possible copper so as to reduce the heating. If the cross-section of the conductor has to be large, it is made more manageable by using cords of twisted copper strands or bundles of copper strips. These latter have the advantage of diminishing the interstices between the consecutive turns. For mean-tension currents the wire is covered with two or three layers of cotton impregnated with varnish or shellac. If the coils are very small, the wire is generally insulated with silk. For high-tension currents the successive layers of wire must be separated by vulcanized fibre, cotton tape impregnated with shellac, Wellesden paper, mica, or ebonite. In special cases the windings are even separated into parts by ebonite partitions perpendicular to the axis of the core, the spaces between the partitions being filled with flat bobbins connected in series. In this manner conductors at very different potentials are kept apart and the danger of disruptive sparks avoided.

If electromagnets are exposed to excessive heating, the wire is insulated with asbestos or mica.

The conductor joining the inner layer of a coil with the outer layer should be made extra strong, for, if it breaks, the coil has to be unwound in order to repair it.

The shape given to the keeper of an electromagnet has a marked influence on the magnet's portative power, which is expressed by $\dfrac{\mathfrak{B}^2}{8\pi} s$ (§ 165). If the polar surface is diminished, the induction $\mathfrak{B}$ is increased, for the lines of force, having a tendency to make their way across the iron rather than across the surrounding air, will go more and more towards this surface in proportion as it grows smaller near the keeper. It follows that the product $\mathfrak{B}^2 s$ will be a maximum for a value of s, generally much smaller than the section of the cores. It is for this reason that keepers are frequently given a convex form in the part adjoining the poles.

166. Magnetization of a Conductor.—When a permanent electric current traverses an iron or steel wire, the molecules situated on the surface of the wire align themselves circularly along the lines of force created by the interior electric flux, forming closed magnetic filaments without external action. To exhibit this transverse magnetization, we need only cut a longitudinal groove in the wire; its edges will exhibit opposite polarities.

167. Modifications in the Properties of Bodies in a Magnetic Field.—A magnetic field causes perturbations in the propagation of light in bodies placed in the field. This discovery, due to Faraday, is proven by the aid of polarized light, whose plane of polarization is altered when it traverses, in the direction of the lines of force, a solid, liquid, or gaseous body placed in a magnetic field.

Faraday used, in his demonstration, an electromagnet whose field-cores are hollowed out along their axis and joined by right-angled iron supports and an iron base.

The substance to be tried is placed between the poles of the magnet. The field-cores are traversed by a ray of light polarized by passing through a Nicol's prism ; on leaving the field-cores the ray is extinguished by an analyzer. When a current is sent through the field-coils the ray reappears :

it can be re-extinguished by giving the analyzer a rotation which measures the angle by which the plane of polarization has turned.

Verdet has shown that this angle is proportional to the difference of the magnetic potentials at the extreme points of the path of the ray through the substance affected. The direction of rotation, however, is independent of the direction of the ray of light, but is different in magnetic and diamagnetic bodies.

This experiment can also be made by the aid of a solenoid traversed by a current and having placed along its axis a tube, closed by glass ends, in which the liquid to be experimented on has been poured. The polarizer is placed at one of the ends of the tube and the analyzer at the other.

If the length of the tube greatly exceeds that of the solenoid, the magnetic potential is practically zero towards its extremities (§ 151); the difference of magnetic potential is equal to the work accomplished by unit pole in moving from one of these extremities to the other, or $4\pi ni$; n denoting the total number of turns, and i the current.

a being a constant, called Verdet's constant, for a given body at a definite temperature, the rotation of the plane of polarization is

$$\theta = a\kappa 4\pi ni.$$

This relation gives the means of measuring the current as a function of the deviation of the polarized ray.

168. Hall Effect.—Hall has discribed a phenomenon which, like the preceding, probably owes its origin to a physical modification in the bodies placed in a field.

If we connect with the poles of a battery the points a and b (Fig. 87) of a thin conductive plate of circular form, we can trace equipotential lines and lines of electric flux distributed as shown in the figure. In order to obtain one of

the equipotential lines, we need only affix at one point of
the plate the extremity of a wire connected with a galvanometer, and move about over the plate, another wire con-

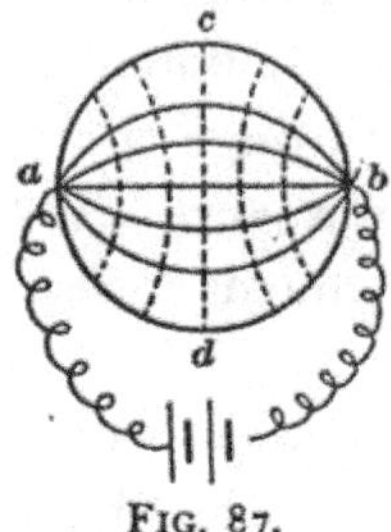

FIG. 87.

nected to the second terminal of the same instrument.
Each time that we meet with a point at the same potential
as the first, the galvanometer will show no deviation.

The equipotential lines, marked by dots, are symmetrical
to the diameter *cd*, whose extremities are at the same potential.

Suppose now that the disc be placed between the poles
of an electromagnet so that the lines of force traverse it
perpendicularly.

The points *c* and *d* immediately cease to be at the same
potential, and a new distribution of the lines of flux is produced as shown in Fig. 88.

FIG. 88.

At first sight it seems as if the displacement of the lines
of flux ought to be attributed to the direct action, governed

by Ampère's rule, of the magnetic force on the electric current.

But experiment shows that the deviations are not in the same direction in different bodies. For example, the deformation of the lines of flux follows Ampère's rule in iron and zinc ; it is the opposite in bismuth and nickel.

At the same time with the above-mentioned displacement, there is observed an apparent increase in the electric resistance of the conductor, attributable to the fact that the mean lines of flux are lengthened by the twisting. This increase of resistance, which is more manifest in conductors of lengthened form placed perpendicularly to the lines of force, can be used, as M. Leduc has shown, to measure the intensity of a magnetic field.

UNITS AND DIMENSIONS.*

GENERAL THEORY.

169. Units.—In every quantitative statement of the value of a physical quantity two factors are involved—the unit of the same concrete kind as the quantity, and a numerical factor. The numerical factor, called the "numeric" for short, is the ratio of the concrete quantity to its unit ; or, it is the number expressing how many times the unit is contained in the concrete quantity. For instance, if l is a definite length and L the unit length, then l/L is the numerical value, or numeric, of this length l.

It is clear that when the unit chosen as a measure of length is changed, the numeric of any concrete length will change inversely ; any length contains twelve times as many inches as feet, the unit being diminished from feet to inches in the ratio of 1 : 12, the numeric increases in the ratio of 12 : 1.

A unit, such as that of length, which involves in its definition no reference to other units, is called a fundamental unit. From the fundamental unit others flow, dependent upon the fundamental units ; these are called derived units.

Such derived units are the unit of area,—the area of a square whose side is the unit length ; the unit of volume, —the volume of a cube whose edge is the unit length ;

---

* By Cary T. Hutchinson, Ph.D.

the unit of velocity,—the velocity with which unit length will be traversed in unit time: these are all simple derived units. Others are more involved, requiring several steps in their reference to the fundamental units. For instance, the unit of acceleration is defined to be the acceleration with which the velocity is increased by unity in unit time; this involves a reference to time and velocity which in turn implies length and time.

The choice of fundamental units is to a great extent arbitrary. Rational systems of units can be built up, starting from many different sets of fundamental units; the units of length, mass, and time are, however, almost universally adopted as fundamental. Practically all physical quantities can be expressed in terms of length, mass, and time.

170. Dimensions.—A change in the fundamental units will obviously change a derived unit. The amount by which a derived unit is affected by given changes in the fundamental units is determined by what are called the "dimensions," or the dimensional equation of the derived unit. These dimensions express the manner in which the fundamental units enter into the derived unit.

A simple illustration of the method of deriving dimensions is that of velocity. Let v, l, and t denote respectively a concrete velocity, length, and time, connected by the relation that the numerical value of the velocity, or its numeric, is equal to the numerical value of the length divided by that of the time; that is, that the given length l would be described in the given time t by a body moving with the given velocity v. Let V, L, T be the units of velocity, length, and time, respectively; then the numerics of v, l, and t are v/V, l/L, t/T, and by assumption,

$$v/V = l/L \div t/T,$$

or

$$v/V = \frac{l}{L} \times \frac{T}{t}.$$

This equation shows that the numeric of the velocity v varies inversely as the unit of length L, and directly as the unit of time T. It also shows that the unit velocity V varies directly as the unit length L, and inversely as the unit time T, since the concrete quantities v, l, and t are not affected by changes in the units V, L, T.

The equation

$$V = L/T = LT^{-1}$$

is known as the dimensional equation of V, and LT^{-1} are the dimensions of V.

It is clear that these dimensions are dependent entirely upon the fundamental units adopted and upon the definition given to velocity; also, that they have no connection with the real physical nature of the quantities to which they refer. As an illustration, suppose force, mass, and time were chosen fundamental units; then since

$$\text{force} = \text{mass} \times \text{acceleration},$$

or

$$f = m \times a = m \times v/t,$$

$$v = f \times t/m,$$

and

$$V = FTM^{-1},$$

an entirely different set of dimensions.

In the same manner the following dimensions are deduced from the ordinary definitions of the quantities involved.

DIMENSIONS OF SOME PHYSICAL QUANTITIES.

Physical Quantity.	Symbol.	Defining Equation.	Dimensions
Length	l		L
Mass	m		M
Time	t		T
Area	s	$s = l^2$	L^2
Volume	V	$V = l^3$	L^3
Velocity	v	$v = l/t$	LT^{-1}
Angular velocity	ω	$\omega = v/l$	T^{-1}
Density		$= m/l^3$	$L^{-3}M$
Force	f	$f = m \times a$	LMT^{-2}
Work	W	$W = f \times l$	L^2MT^{-2}
Power	P	$P = W/t$	L^2MT^{-3}
Pressure	p	$p = f/s$	$L^{-1}MT^{-2}$
Moment of inertia	K	$K = Ml^2$	L^2M

The other mechanical and dynamical quantities can readily be deduced from these in the same manner.

171. C. G. S. System of Units.—The system of units adopted by the entire scientific world is that based upon the centimetre as the unit length, the gramme as the unit mass, and the second as the unit time.

These units were chosen after long consideration because:—they admit of accurate comparison with quantities of their own kind, the units can be determined easily and accurately and standardized at all times and places, and they lead to simple definitions of the most important physical quantities, with their relations to one another.

In all that follows L will denote unit length or one centimetre, M the unit mass or one gramme, and T the unit time or one second.

The C. G. S. system of units applies to mechanical as

well as electrical quantities, although it has been given prominence mainly through its electrical applications.

ELECTRICAL AND MAGNETIC UNITS.

172. General Considerations.—A consistent system of electrical and magnetic units can be devised with any one of the equations defining an electric or magnetic relation as a starting-point. Each system so formed would have different dimensions for the same concrete quantity, and these dimensions, as explained above, would serve to show the changes in the various units when the fundamental units were changed, and consequently the ratio of the various units in the different systems.

Two systems have been built up in this way, and are in use to-day: the first, the Electrostatic System (E. S.), is based on the definition of unit quantity of electricity; the second, the Electromagnetic System (E. M.), is based on the definition of unit magnetic pole. In order that magnetic quantities may be expressed in terms of the electrostatic units, and electric quantities in terms of electromagnetic units, some connecting link between the two sets of phenomena must be used: this link is the relation developed in § 136, that the magnetic force due to an unlimited rectilinear current i at a point at a distance l, is proportional to the current i, and inversely proportional to the distance l. That is,

$$\text{Magnetic force} = a \times i/l.$$

If $\mathcal{H}$ is the unit magnetic force and I the unit current,

$$\mathcal{H} = IL^{-1} \quad \ldots \quad \ldots \quad \ldots \text{(A)}$$

There is only one known relation connecting electric current and magnetic force; equation (A) is one expression of this relation. Another common form of expressing the same relation is,—the work done by unit pole in passing around any closed curve that threads the electric circuit equals $4\pi i$. These two relations, and others of the same kind, are interchangeable.

Coulomb's experiments connect mechanical force with electric quantity ; as ordinarily stated,

$$f = q^2/l^2.$$

This statement assumes the action to take place in air (or *vacuo*), and does not bring the medium into evidence ; for greater generality the expression should be

$$f = q^2/(l^2K), \quad . \quad . \quad . \quad . \quad . \quad . \quad \text{(B)}$$

where K is the specific inductive capacity or permittivity of the medium.

Similarly, for magnetic quantities,

$$f = m^2/(l^2\mu), \quad . \quad . \quad . \quad . \quad . \quad . \quad \text{(C)}$$

where μ is the permeability of the medium.

173. Systems in Terms of K and μ.—Starting with equations (A) and (B), a system of units can be developed in which all the electric and magnetic quantities are expressed in terms of LMT and K. Similarly, starting with (A) and (C), a system can be developed in which all the quantities are expressed in terms of LMT and μ.

These two systems are the only ones in use : the first, the system in terms of K, leads directly to the Electrostatic System : the second, the system in terms of μ, leads to the Electromagnetic System. The two are herein developed side by side ; the quantities are chosen in pairs so as to show, as far as possible, the similarity of each in the K-system to its fellow in the μ-system. Units in terms of K will be distinguished by a line over the symbol.

Electric Quantity (q) :

In terms of K.—Unit quantity is defined by equation (B); that is, the mechanical force f acting between two equal quantities q at distance l is

$$f = q^2/(l^2 K),$$

$$\overline{Q}^2 = FL^2 K$$

$$= LMT^{-2}L^2 K,$$

$$\overline{Q} = L^{\frac{3}{2}}M^{\frac{1}{2}}T^{-1}K^{\frac{1}{2}}.$$

Electric Current (i) :

In terms of K.—Current is defined as the quantity of electricity carried across the section of a conductor in unit time ; or,

$$i = q/t,$$

$$\overline{I} = \overline{Q}T^{-1}$$

$$= L^{\frac{3}{2}}M^{\frac{1}{2}}T^{-2}K^{\frac{1}{2}}.$$

Strength of Magnetic Pole (m) :

In terms of μ.—Pole strength is defined by equation (C), similar to equation (B) for quantity. Therefore,

$$m = L^{\frac{3}{2}}M^{\frac{1}{2}}T^{-1}\mu^{\frac{1}{2}}.$$

Magnetic Force ($\mathcal{H}$) :

In terms of μ.—Magnetic force is defined as the force acting on unit pole ; therefore,

$$\mathcal{H} = F/m$$

$$= L^{-\frac{1}{2}}M^{\frac{1}{2}}T^{-1}\mu^{-\frac{1}{2}}.$$

Electric Current (i):

In terms of μ.—Equation (A), the linking equation, gives the magnetic force due to unit current i at distance l as

$$\mathfrak{H} = a \times i/l,$$

$$I = \mathfrak{H}L$$

$$= L^{\frac{1}{2}}M^{\frac{1}{2}}T^{-1}\mu^{-\frac{1}{2}}.$$

Electric Quantity (q):

In terms of μ.—

$$q = i \times t,\ \text{as above}$$

$$Q = I \times T$$

$$= L^{\frac{1}{2}}M^{\frac{1}{2}}\mu^{-\frac{1}{2}}.$$

Magnetic Force (𝔥):

In terms of K.—

$$\mathfrak{H} = i/l,$$

$$\overline{\mathfrak{H}} = \overline{I}/L$$

$$= L^{\frac{1}{2}}M^{\frac{1}{2}}T^{-1}K^{\frac{1}{2}}.$$

Strength of Magnetic Pole (m):

In terms of K.—

$$f = \mathfrak{H}m,\ \text{as above};$$

$$\overline{m} = F\mathfrak{H}$$

$$= L^{\frac{1}{2}}M^{\frac{1}{2}}K^{-\frac{1}{2}}.$$

All the electric magnitudes depend on q, from which they are easily derived, following the usual definitions; hence, having q expressed both in terms of K and of μ, the expression for all the electric magnitudes can be deduced in terms, both of K and of μ. Similarly, having pole strength m, expressed both in terms of K and of μ, the expression for all the other magnetic magnitudes can be deduced both in terms of K and μ.

The accompanying table gives the symbol, the defining equation, and the dimensions, both in terms of K and of μ, of all the common electrical and magnetic quantities; it gives also the ratio of the dimensions in terms of K to the dimensions in terms of μ.

In the table, the symbols and names recommended by the Committee on Notation of the Chamber of Delegates of the International Electrical Congress of 1893 are used, this system having been adopted throughout the book.

174. Proposed Nomenclature.—The table contains the names and symbols of the quantities in common use; names have been proposed for the ratios of Flux/Force and Flux-Density/Force-Intensity for the dielectric flux to correspond to those already in use for the conductive and magnetic fluxes.

For the conductive flux :

Flux/Force $= i/e =$ Conductance $(g) =$ 1/Resistance (r).

Flux-Density/Force-Intensity $= i/l^2 \div e/l =$ Conductivity $(\gamma) =$ 1/Resistivity (ρ).

For the magnetic flux :

Flux/Force $= \Phi/\mathcal{F} =$ Permeance $=$ 1/Reluctance $(\mathcal{R})$.

Flux-Density/Force-Intensity $= \mathcal{B}/\mathcal{H} =$ Permeability (μ) $=$ 1/Reluctivity (ν).

For the dielectric flux two systems of names have been proposed, the first by Oliver Heaviside and the second by F. E. Nipher. Heaviside's * notation is as follows :

Flux/Force $= N/e =$ Permittance $=$ 1/Elastance.

---

* *Electromagnetic Theory*, Vol. I.

TABLE OF ELECTRIC AND MAGNETIC QUANTITIES.

Physical Magnitude.	Symbol.	Defining Equation.	L	M	T	K	L	M	T	μ	Ratio of $\dfrac{K\text{-system}}{\mu\text{-system}}$
Electric quantity	q	$f = q^2/l^2 K$	3/2	1/2	$-$1	1/2	1/2	1/2	$\cdot$0	$-$1/2	$V(K\mu)^{\frac{1}{2}}$
Pole strength	m	$f = m^2/l^2\mu$	1/2	1/2	0	$-$1/2	3/2	1/2	$-$1	1/2	$V^{-1}(K\mu)-\frac{1}{2}$
Electric force[1]	F	$F = f/q$	$-$1/2	1/2	$-$1	$-$1/2	1/2	1/2	$-$2	1/2	$V^{-1}(K\mu)-\frac{1}{2}$
Magnetic force[2]	$\mathcal{H}$	$\mathcal{H} = f/m$	1/2	1/2	$-$2	1/2	$-$1/2	1/2	$-$1	$-$1/2	$V(K\mu)^{\frac{1}{2}}$
Electromotive force[3]	e	$e = lF$	1/2	1/2	$-$1	$-$1/2	3/2	1/2	$-$2	1/2	$V^{-1}(K\mu)-\frac{1}{2}$
Magnetomotive force[4]	$\mathcal{F}$	$\mathcal{F} = l\mathcal{H}$	3/2	1/2	$-$2	1/2	1/2	1/2	$-$1	$-$1/2	$V(K\mu)^{\frac{1}{2}}$
Displacement[5]	D	$D = KF$	$-$1/2	1/2	$-$1	1/2	$-$3/2	1/2	0	$-$1/2	$V(K\mu)^{\frac{1}{2}}$
Induction[6]	$\mathcal{B}$	$\mathcal{B} = \mu\mathcal{H}$	$-$3/2	1/2	0	$-$1/2	$-$1/2	1/2	$-$1	1/2	$V^{-1}(K\mu)-\frac{1}{2}$
Electric flux	N	$N = l^2 D$	3/2	1/2	$-$1	1/2	1/2	1/2	0	$-$1/2	$V(K\mu)^{\frac{1}{2}}$
Magnetic flux	Φ	$\Phi = l^2\mathcal{B}$	1/2	1/2	0	$-$1/2	3/2	1/2	$-$1	1/2	$V^{-1}(K\mu)-\frac{1}{2}$
Specific inductive capacity	K	—	0	0	0	1	$-$2	0	2	$-$1	$V^2(K\mu)$
Permeability	μ	—	$-$2	0	2	$-$1	0	0	0	1	$V^{-2}(K\mu)-1$
Permeance	—	$= \Phi/\mathcal{F}$	$-$1	0	2	$-$1	1	0	0	1	$V^{-2}(K\mu)-1$
Reluctance	$\mathcal{R}$	$\mathcal{R} = \mathcal{F}/\Phi$	1	0	$-$2	1	$-$1	0	0	$-$1	$V^2(K\mu)$
Reluctivity	ν	$\nu = \mathcal{R}l$	0	0	$-$2	1	$-$2	0	0	$-$1	$V^2(K\mu)$
Current	i	$i = q/t$	3/2	1/2	$-$2	1/2	1/2	1/2	$-$1	$-$1/2	$V(K\mu)^{\frac{1}{2}}$
Resistance	r	$e = ir$	$-$1	0	1	$-$1	1	0	$-$1	1	$V^{-2}(K\mu)-1$
Resistivity	ρ	$\rho = rl$	0	0	1	$-$1	2	0	$-$1	1	$V^{-2}(K\mu)-1$
Conductance	g	$g = 1/r$	1	0	$-$1	1	$-$1	0	1	$-$1	$V^2(K\mu)$
Conductivity	γ	$\gamma = g/l$	0	0	$-$1	1	$-$2	0	1	$-$1	$V^2(K\mu)$
Inductance[7]	L	$\Phi = Li$	$-$1	0	2	$-$1	1	0	0	1	$V^{-2}(K\mu)-1$
Capacity	c	$q = ce$	1	0	0	1	$-$1	0	2	$-$1	$V^2(K\mu)$

[1] Electric Field Strength or Intensity ; Electromotive Intensity.
[2] Magnetic Field Strength or Intensity; Magnetizing Force.
[3] Electric Potential ; Voltage.
[4] Line Integral of Magnetic Force ; Magnetic Potential.
[5] Electric Induction ; Electric Flux-Density ; Faraday Tubes.
[6] Magnetic Flux-Density ; Density of Lines of Magnetic Force.
[7] Coefficient of Induction ; Electromagnetic Capacity.

Flux-Density/Force-Intensity $= D/F =$ Permittivity $=$ 1/Elasticity.

Permittance is capacity; Permittivity is specific inductive capacity.

Nipher's * notation is as follows:

Flux/Force $= N/e =$ Perviance $=$ 1/Diviance.

Flux-Density/Force-Intensity $=$ Perviability $=$ 1/Diviability.

Perviance is capacity; perviability is specific inductive capacity.

Neither of these last sets of names is in use.

175. Dimensions of K and μ.—The table shows that the dimensions of the unit of every quantity in the K-system are related to its dimensions in the μ-system by a simple function of $LT^{-1}(K\mu)^{\frac{1}{2}}$, in which M does not enter, namely,— the quantity, its reciprocal, its square, and the reciprocal of its square. The dimensions thus deduced are entirely indeterminate so long as K and μ have no dimensions in LMT. To give K and μ, coefficients depending on the nature of the medium, dimensions in LMT, implies a mechanical theory of electricity and magnetism.

If the phenomena of electricity and magnetism are different manifestations of mechanical action of one kind or another, then the quantities K and μ must have dimensions in LMT, and the expression of the dimensions of any quantity, as q, in these two systems must be identical; that is, their ratio, a function of $LT^{-1}(K\mu)^{\frac{1}{2}}$ must be zero dimensions, or $L^{\circ}M^{\circ}T^{\circ}$; or,

$$(K\mu)^{-\frac{1}{2}} = LT^{-1} = v, \text{ a velocity.}$$

Maxwell, in his electromagnetic theory, which does assume all the phenomena of electricity and magnetism to be

---

manifestations of ordinary mechanical forces, shows that the velocity of propagation of electric disturbances in a medium of specific inductive capacity, K, and permeability, μ, is $(K\mu)^{-\frac{1}{2}}$.

The ratio of the dimensions being in all cases a function of $V(K\mu)^{\frac{1}{2}}$, any values of K and μ that make this expression unity will result in identical dimensions in the two systems for every quantity. Any values of K and μ that are based on the assumption of a mechanical theory of electricity will fulfil this condition.

Many different systems of dimensions have been proposed of late, all based on certain assumptions for the values of K and μ; the number that can be devised is without limit, but as there is no reality behind the assumptions it is useless to record them.

The two systems of units ordinarily met with are the Electrostatic (E. S.) and the Electromagnetic (E. M.): the first is based on the assumption that the specific inductive capacity of vacuum is unity, and is therefore deduced from the K-system by making K unity in all the formulæ; the second is based similarly on the assumption that the permeability of vacuum is unity, and is therefore deduced from the μ-system by making μ unity in all the formulæ.

These two systems are the only ones that are in use, and of these the Electrostatic system has no practical importance; all electrical measurements as ordinarily made are in terms of the Electromagnetic system, or in the " Practical System," of· which the units are merely multiples of the units on the Electromagnetic system.

The values of unity for both K and μ were chosen merely for simplicity; they assume vacuum as the standard medium, and ignore the physical character of the two quantities; that is, they assume them to be merely numerical coefficients. As simultaneous values, they do not satisfy the condition $(K\mu)^{-\frac{1}{2}} =$ velocity, and hence do not lead to identical dimensions.

176. Value of Ratio "v."—The ratio of the dimensions of any unit in the Electrostatic system to its dimensions in the Electromagnetic system—that is, the function of $V(K\mu)^{\frac{1}{2}}$—now reduces to a function of "v"; this "v" is a definite concrete velocity, and its numerical value can readily be found by determining the value of any electric or magnetic quantity, first in one system and then in the other, the ratio of the numerical values representing a function of "v." This velocity "v" has been determined experimentally by comparing the two measures of many of the electric units. The best experiments give for the value of "v" very nearly 3×10^{10} centimetres per second,—the same as the value of the velocity of propagation of light *in vacuo*. As "v," on Maxwell's theory, is the velocity of propagation of an electric disturbance, these disturbances are then propagated with the velocity of light. But, all theory aside, experiment shows this ratio "v" to be practically the velocity of light, whatever it may represent physically.

There are a number of ways in which the physical significance of "v" can be illustrated; one of the simplest is shown in the following ideal experiment :

Assume a conductive sphere, which is at the same time compressible; the capacity of a sphere is equal in E. S. measure to its radius. That is,

$$q = l \times e,$$

where l is the radius and e the potential to which it is charged.

Suppose this sphere to be connected to the ground through a resistance of which the value in E. S. units is r, the sphere will then discharge and its potential will fall; but imagine the sphere to be compressed at the same time at such a rate that the increase of potential due to compression equals the diminution due to the discharge.

Then as e is constant, a constant current will flow, represented by

$$i = e/r.$$

But

$$q = l \times e,$$

and

$$\frac{dq}{dt} = i = \frac{edl}{dt} = e/r.$$

That is,

$$r = 1 \div dl/dt.$$

Or, the velocity dl/dt with which the radius of the sphere must be shortened is numerically equal to the conductance of the discharge circuit in E. S. units.

The chief use of these dimensions is to show the manner in which the fundamental units enter into the composition of the unit of any quantity, and thus show the changes in the derived unit flowing from given changes in the fundamental units. For instance, if LMT are changed to $L'M'T'$, the new E. S. unit of quantity will be equal to

$$\left(\frac{L'}{L}\right)^{\tfrac{3}{2}}\left(\frac{M'}{M}\right)^{\tfrac{1}{2}}\left(\frac{T'}{T}\right)^{-}$$

times the old E. S. unit; similarly the new E. M. unit of quantity will be equal to

$$\left(\frac{L'}{L}\right)^{\tfrac{1}{2}}\left(\frac{M'}{M}\right)^{\tfrac{1}{2}}$$

times the old E. M. unit.

177. Practical System.—In the E. M. system the units of the quantities in common use are either impracticably large or small for convenient use.   This fact had led to the adoption and use of a so-called " Practical " system, based on a unit of length equal to 10^9 centimetres, or an earth's quadrant ; a unit of mass equal to 10^{-11} grammes, and a unit of time equal to the second.   That is, in the Practical system,

$$L' = 10^9 \text{ centimetres,}$$

$$M' = 10^{-11} \text{ grammes,}$$

$$T' = 1 \text{ second.}$$

From the table of dimensions of the various quantities the accompanying table is derived, showing the relation between the values of the various units in this Practical system and in the C. G. S., Electromagnetic system, for electric and magnetic quantities.

RELATION BETWEEN PRACTICAL AND C. G. S. ELECTRO-MAGNETIC SYSTEMS.

Quantity.	Name of Unit in Practical System.	Practical Unit equals
Length	Quadrant	10^9 centimetres
Mass		10^{-11} grammes
Time	Second	10^0 seconds
Area		10^{18} $\overline{\text{cm}}^2$
Velocity	....	10^9 cm/s
Force	....	10^{-2} dyne
Work	Joule	10^7 erg
Power	Watt	10^7 erg/s
Resistance	Ohm	10^9 cm/s
Current	Ampere	10^{-1} C. G. S.
Quantity	Coulomb	10^{-1} C. G. S.
Electromotive force	Volt	10^8 C. G. S.
Magnetic flux	....	10^8 Weber
Magnetic flux-density	....	10^{-10} Gauss
Reluctance	....	10^{-9} Oersted

Quantity.	Name of Unit in Practical System.	Practical Unit equals
Permeance..............		10^9 C. G. S.
Magnetic force....		10^{-10} C. G. S.
Magnetomotive force		10^{-1} Gilbert
Capacity...	Farad	10^{-9} C. G. S.
Inductance............	Henry	10^9 C. G. S.
Specific inductive capacity		10^{-18} C. G. S.

178. Nomenclature of Practical Units. — The more common units in the Practical system have long had the names of distinguished men of science, such as Ohm, Volta, Ampère, Coulomb, Joule, Watt, Faraday. These names have answered all ordinary purposes until the last few years; but now a demand has arisen, owing to the more common use of the quantities, for names for the units of flux, induction, magnetomotive force, reluctance, inductance, and some others. The matter has been discussed widely at electrical congresses and before electrical societies.

The units of the Practical system were selected to be of convenient size for the common electric quantities; they serve fairly well for this purpose, with the exception of the Farad, which is about one million times too large, and which is replaced in practice by the microfarad. But the Practical system does not give magnetic units of convenient magnitude; for instance, the unit of flux would be 10^8 C. G. S. units, which is about one hundred times too large; the unit of flux-density or induction would be 10^{-10} C. G. S. units, whereas the C. G. S. unit itself is more nearly right; the unit of reluctance is 10^{-9} C. G. S. units, whereas the C. G. S. unit is again approximately right. The latest official body to pass upon the question, the International Electrical Congress of 1893, refused to adopt new magnitudes or names for the magnetic units, but recommended the adoption of the C. G. S. system. This same congress, however, adopted the Henry as the unit of induction in the Practical system, and defined it as follows : " As a unit of induction the henry is recommended, which is the induction in a circuit when the electromotive force induced in this circuit is one international volt, while the inducing current varies at the rate of one ampere per second." It may be considered official.

Since this date, however, the American Institute of Electrical Engineers has adopted, " provisionally," the names Weber, Gauss, Gilbert, and Oersted for the C. G. S. units of flux, flux-density, magnetomotive force, and reluctance, respectively, as shown by the table. It adopts Gauss for the unit of magnetic force as well as for magnetic flux-density; these two quantities are different physical magnitudes, for which the same name should not be used.

These names have not been generally accepted, and should be considered merely provisional. The need for such names is in many cases not clear; the giving of names of individuals to C. G. S. units is not in conformity with previous usage in the matter.

The British Association has recently proposed a different system of names and magnitudes. For unit flux it recommends 10^8 C. G. S. units, and calls it a Weber, thus agreeing in name only with the American Institute of Electrical Engineers ; for unit magnetomotive force, to be called the Gauss, two magnitudes are suggested, one, $4\pi ni$, and the other, ni, or the ampere-turn. There are other recommendations of minor importance.

179. " Rational " System.—Another system of units differing fundamentally from these has been developed by Oliver Heaviside, and christened by him the " Rational " System.

The underlying assumption in the E. S. and E. M. systems is that the action takes place directly between unit quantities of electricity or unit poles of magnetism; they ignore the existence of the medium entirely. Heaviside's system is based on the assumption that all the phenomena are manifestations of stresses and strains in a medium; the medium is therefore brought into evidence. This plan eliminates the factor 4π from the expression of magnetomotive force, making it correspond to ampere-turns; this is merely an incidental advantage of the " Rational" system, but it has been seized upon as the main reason for its adoption.

Coulomb's law for the force existing between two quantities m and m' at distance l,

$$f = mm'/(\mu l^2),$$

or, in the E. M. system,

$$f = mm'/l^2, \quad \mu = 1,$$

Heaviside puts

$$\mathcal{H} = m/(4\pi l^2),$$

and

$$f = \mathcal{H}m' = mm'/4\pi l^2.$$

That is, instead of making the flux from a pole of strength m equal to $4\pi m$, he makes it equal to m.

The question is simply, What is the rational or natural method of measuring the strength of a source at a given distance from its centre? Coulomb's formula follows the ordinary astronomical or gravitational method; but this is clearly illogical: the strength of a source m at a distance l is the quantity of m per unit surface at that point. The total flux from the source is necessarily, by the principle of continuity, the quantity at the source. Its strength at distance l, when the law of variation is that of the inverse square, is $m/4\pi l^2$, or the strength of the field; the force which the field exerts on another source m' is equal to the product of m' by the strength of the field at the point,—that is,

$$mm'/4\pi l^2.$$

This simply means in ordinary language that unit pole sends out one line of force, not 4π lines.

A similar irrationality would be introduced into the geometrical measures if unit area were defined to be the area of a circle whose diameter is unity, instead of the area of a square whose side is unity. If this were done, the expression for areas and volumes of rectangular surfaces and solids would involve 4π as a factor, while 4π would be eliminated from the expression for area of circle and volume of sphere, where they properly belong.

The changes that this makes in the various units are readily obtained by giving to μ the value 4π in the table showing the dimensions in the μ-system.

Denoting units in the Rational system by the subscript "r," then these relations are found:

$$(4\pi)^{\frac{1}{2}}=e_r/e=i/i_r=q/q_r=\mathcal{K}/\mathcal{K}=\mathcal{B}_r/\mathcal{B}=E_r/E=\Phi_r/\Phi=\mathcal{F}/\mathcal{F}_r.$$

$$(4\pi) = r_r/r = L_r/L = \mathcal{R}/\mathcal{R}_r = c/c_r.$$

Furthermore, $1 = W_r/W = P_r/P$, since the expression for energy does not involve μ.

This gives the relations of the C. G. S. units.

In passing to a new Practical system, either the same or new powers of ten could be used. Heaviside seems to prefer the use of a single multiple, 10^8 or 10^{-8}; this, however, has the great disadvantage of changing the units of energy and power; it seems therefore preferable to adhere to the same factors in passing from the Rational C. G. S. to the new Practical system. On this assumption,

$$\begin{aligned}
\text{``Rational'' Ohm} &= 4\pi \times \text{Ohm} \\
\text{`` \quad Ampere} &= (4\pi)^{-\frac{1}{2}} \times \text{Ampere} \\
\text{`` \quad Volt} &= (4\pi)^{\frac{1}{2}} \times \text{Volt} \\
\text{`` \quad Watt} &= \text{Watt} \\
\text{`` \quad Weber} &= (4\pi)^{\frac{1}{2}} \times \text{Weber} \\
\text{`` \quad Gilbert} &= (4\pi)^{-\frac{1}{2}} \times \text{Gilbert} \\
\text{``} &= \text{Rational Ampere-turn}
\end{aligned}$$

There has been considerable agitation recently in England looking to the introduction of these units into practical use. The changes suggested are great, but not nearly so radical as were those involved in the introduction of the decimal system. The numerical factors would prove troublesome in workshop practice for a time, but this would soon pass away, and a simpler and better system, and, above all, a logical one, would remain.

180. Electrical Standards of Measure.—The various units have all been defined in terms of the C. G. S. system ; the physical properties on which these definitions rest are such as do not readily lend themselves to physical measurements ; concrete standards based directly on these definitions would not be capable of quick and easy reproduction and of ready comparison at all places and times. Hence certain concrete standards that do meet these conditions have been chosen for some of the units ; these have been measured in terms of C. G. S. units, by the most accurate and refined methods and at the hands of the most careful experimenters.

Resistance.—In the C. G. S. E. M. system resistance is defined as the ratio of electromotive force to current, both of which in turn are referred back to the definition of unit pole ; the concrete standard of resistance has been a mercury column of certain length and cross-section; this was first used in the form of the Siemens Unit, as a more or less arbitrary standard; the value given to the length, 100 cm., was not intended to reproduce the ohm exactly.

The Electrical Congress of 1881 defined the ohm to be the resistance at 0° C. of a column of mercury of uniform cross-section of one square millimetre, but left the exact length to be fixed by subsequent experiment. Many experimental researches were made to fix this length, all giving values approximately 106 cm. So the Paris Congress of 1884 adopted this value under the name Legal Ohm. The British Association in 1886 endorsed this value, to hold for a period of ten years. Later and more accurate measurements have, however, shown that this value is too small and that the correct value is 106.3 cm., to within about $\frac{1}{5000}$. This value was adopted by the International Congress of 1893 at Chicago under the name International Ohm.

An earlier standard, adopted by the British Association in 1864, is the British Association Unit, abbreviated to B. A. U.

The values of these various units in terms of mercury are as follows:

Siemens Unit...................... 100 cm.
B. A. U........................... 104.8
Legal Ohm........ 106
International Ohm 106.3

In terms of the International Ohm:

Siemens Unit........ ...... .9408 Int. Ohm.
B. A. U.............. ...... .9863 Int. Ohm.
Legal Ohm............ ...... .9972 Int. Ohm.
Int. Ohm... .... 1.000 Int. Ohm.

Current.—Unit current is the Ampere, defined to be one tenth of the C. G. S. unit; its concrete representation is obtained in the form of a certain weight of silver deposited by the current in a silver voltameter, constructed in accordance with certain definite specifications.

The most accurate determination of this constant gives the weight of silver deposited by one ampere in one second as 1.118 milligrammes. This definition was adopted by the International Congress of 1893, under the name International Ampere.

Electromotive Force.—The unit is the Volt, defined to be 10^8 C. G. S. units; it is represented by the electromotive force of a certain form of Clark cell prepared in accordance with certain specifications. The $\frac{1000}{1434}$ part of the electromotive force of this cell, at a temperature of $15°$ C., was adopted by the 1893 Congress under the name International Volt.

181. Recommendations of Congress of 1893. — The same Congress adopted these definitions:

International Coulomb, the quantity of electricity transferred by an International Ampere per second.

International Farad, the capacity of a condenser charged to one International Volt by one International Coulomb.

Joule, the unit of work, equal to 10^7 C. G. S. units and represented by the energy expended in one second by one International Ampere on one International Ohm.

Watt, the unit of power, equal to 10^7 C. G. S. units of power and represented by the work done at the rate of one joule per second.

Henry, the unit of induction, the induction in a circuit when the electromotive force induced in the circuit is one International Volt, while the inducing current varies at the rate of one International Ampere per second.

ELECTROMAGNETIC INDUCTION.

LAWS OF ELECTROMAGNETIC INDUCTION.

182. Induced Currents.—Faraday discovered, in 1831, a series of phenomena, called phenomena of *electromagnetic induction*, consisting in the production of currents in conductors which are displaced in a magnetic field or placed in a variable magnetic field.

The circumstances which give rise to *induction currents* may be summed up as follows:

I. Suppose two neighboring circuits a and b, such as, e.g., two concentric solenoids. When a current starts in a, a current in the opposite direction passes through b.

The first current is called the *inducing*, the second the *induced*, current: when the inducing current has reached a constant strength, the induced current ceases. If the inducing current gradually diminishes to zero, a new induced current is produced in the circuit b, this time in the same direction as the inducing current and ceasing with it. These two induced currents are respectively called the inverse-secondary or make-induced current, and the direct-secondary or break-induced current.

Now, bring near to a conductor b a conductor a, traversed by a constant current; b becomes the seat of an inverse-secondary current which lasts as long as the movement. If we move a away, a direct-secondary current arises in b.

To sum up, there is an induced current every time and

as long as the strength or position of the inducing current varies.

II. Since a magnet can be compared to a solenoid, it is easy to see that every variation of position or magnetization of a magnet creates induced currents in a neighboring circuit, the direction of which is determined by substituting for the magnet an equivalent solenoid whose position or current experiences corresponding variations.

III. Every variation in the form of a circuit or of its current gives rise to induction phenomena.

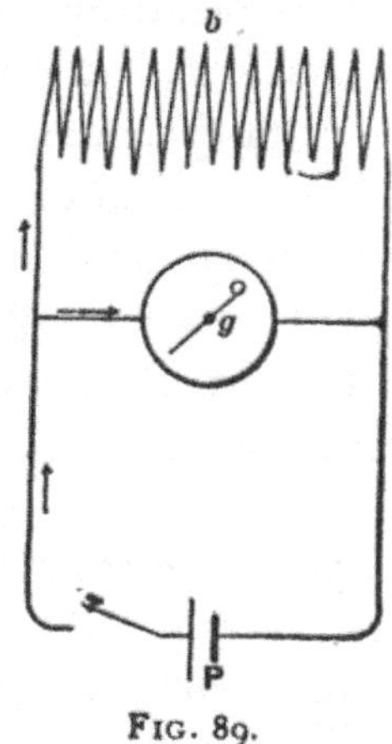

FIG. 89.

Suppose a circuit containing a solenoid b, with or without an iron core, Fig. 89: a galvanometer inserted in a shunt, g, shows a fixed deviation, equal to an angle α, when the circuit is closed by a switch.

Suppose we keep the needle at this angle by a stop which prevents it from returning to zero, and after having broken the circuit, close it again. The needle is thrown suddenly beyond the angle α, which indicates a current superadded to the normal one in the galvanometer, and which may be considered as the inverse-induced current of the solenoid.

Now bring the needle back to zero and keep it there, while the current is passing, by means of a stop on the side towards which it tends to swing.

At the moment of breaking the circuit, we shall observe an impulse in the direction opposite to a, attributable to a direct-induced current developed in the solenoid.

These induction effects, which are very marked when the solenoid has a large number of turns, and especially when furnished with an iron core, are called *inverse extra-current* or *inverse-induced current* and *direct extra-current* or *direct-induced current*. These extra-currents are clearly shown in their physiological effects: e.g. the difference of potential produced by a single voltaic element is insensible to the touch; but if a circuit be formed containing a cell and an electromagnet, and if a person touch the two terminals of the latter with his fingers, he feels a slight shock when the circuit is broken, due to part of the break-induced extra-current passing across the hand.

183. Lenz' Law.—Upon analyzing these various orders of phenomena, we observe that every circumstance which brings about an induction current also brings about a modification of the flux of magnetic force across the circuit acted on.

The case of induction by displacement of a circuit in a magnetic field can be considered as the counterpart of the displacement of a current under the action of a field. This analogy has been investigated by Lenz, who has given the following law for the induction produced by movement.

Every relative displacement of a circuit and a field generates an induced current which tends to oppose the displacement by means of its electromagnetic reaction.

Thus, when a conductor is displaced in a field, e.g. towards the left of a figure lying along the conductor and facing in the direction of the lines of force, the induced current goes from the head of the figure to its feet; the current that would produce the same displacement would flow from its feet to its head.

184. General Law of Induction.—Lord Kelvin and Helmholtz have deduced from the principle of the conservation of energy a law which gives in every case the direction and strength of the induced current.

Suppose a circuit, in a magnetic field, contains a cell whose electromotive force is e; call the total resistance of the circuit r; the normal current is, by Ohm's law,

$$I = \frac{e}{r}.$$

Under these conditions the energy of the current is transformed into heat, so that in a time dt

$$eIdt = I^2 r dt.$$

But in consequence of the reaction of the field on the current, an electromagnetic force is set up which tends to displace the current and make it do work, dW, during the time dt. This work is naturally accomplished at the expense of the cell's energy, being the only source of energy at hand; and as e and r are constants by hypothesis, it follows that the current has had to assume a new value i such that

$$eidt = i^2 r dt + dW.$$

But we have seen (§ 142) that the electromagnetic work is equal and of contrary sign to the variation in the energy, $-i\Phi$, of the current in the field. Consequently

$$dW = -dw = id\Phi;$$

therefore

$$eidt = i^2rdt + id\Phi,$$

whence

$$i = \frac{e - \dfrac{d\Phi}{dt}}{r}. \quad . \quad . \quad . \quad . \quad . \quad . \quad . \quad (1)$$

We see that the term $-\dfrac{d\Phi}{dt}$ plays the part of an electro-motive force which tends to diminish the current; it is the general expression for the *electromotive force of induction.*

If we discard the cell and then displace the circuit in the same way as before, by an expenditure of mechanical en-ergy, an electromotive force of induction, $-\dfrac{d\Phi}{dt}$, is set up, which produces a current

$$i = \frac{-\dfrac{d\Phi}{dt}}{r}. \quad . \quad . \quad . \quad . \quad . \quad . \quad . \quad (2)$$

Equation (2) shows that an immovable circuit traversed by a variable magnetic flux is equally the seat of an *electro-motive force of induction which is in every case equal and of contrary sign to the rate of variation of the flux with regard to the time.*

185. Maxwell's Rule.—The direction of the electromo-tive force is readily deduced from Maxwell's rule (§ 135). If we suppose a corkscrew to move forward, rotating in the direction of the field, the direction of rotation indicates the positive direction of the electromotive force. Now, when the flux or the number of lines of force decreases, the elec-tromotive force $-\dfrac{d\Phi}{dt}$ is positive. If the flux increases, the electromotive force is negative, i.e. it is in the opposite direction to the movement of rotation of the corkscrew.

To sum up: *the electromotive force is in the direction of the rotation of a corkscrew advancing in the direction of the lines of force, when the flux decreases; when the flux increases, it is in the opposite direction.*

Apply this rule to a solenoid into which a magnet enters by its *N* end. The flux of force across the solenoid is increased, consequently the electromotive force is negative. When the magnet is drawn out, the flux decreases and the electromotive force becomes positive.

186. Faraday's Rule.—In the case of induction by displacement, it is often useful to determine the electromotive force due to the movement of the various parts of the circuit acted upon. Now it will be remembered (§ 145) that the total variation of the flux across a circuit is the algebraic sum of the elementary fluxes cut by its various parts.

We can, therefore, interpret the general law of induction by saying that *the electromotive force set up in a conductor is measured at each instant by the flux of force cut in unit time.* In other words, the electromotive force equals the length of the conductor multiplied by the field-intensity, and by the projection normal to the lines of force of the displacement made in unit time.

The direction of the electromotive force is obtained by Lenz' law (§ 183).

Thus the axle of a car, which, in the northern hemisphere, moves from east to west cutting the terrestrial lines of force, is the seat of an electromotive force of induction directed from north to south. To find this direction, we need only suppose a figure lying along the axle and looking downwards (since in the northern hemisphere the lines of force go downwards), and that the figure is displaced to the left; the induced current then tends to pass from the head to the feet.

Faraday's rule has the advantage over Maxwell's of apply-ing even in the case of an open circuit, and of showing that an electromotive force of induction exists in conductors cutting the lines of force of a magnetic field.

Fleming's mnemonic rule can also be used (§ 138).   In the present case the right hand should be used : The index finger and thumb being respectively in the direction of the lines of force and the displacement, the middle finger will indicate the direction of the induced E. M. F.

It follows from the above rules that if a circuit is dis-placed in a field in such a way that the flux embraced by it remains constant, or when a conductor moves parallel to the lines of force in a field, no induced currents are gener-ated.

187. Seat of the Electromotive Force of Induction.— The electromotive force takes its rise in all the parts of the circuit which cut lines of force, and in these only.   Thus, in the above example of an axle moving upon rails, the E. M. F. is developed in the axle.

Suppose that a magnetized bar be displaced along the axis of a metallic ring.   All the elements of the ring cut in each instant the same number of lines of force, and in each element, with a resistance dr, is set up an E. M. F. de and a current

$$i = \frac{\mathrm{d}e}{\mathrm{d}r}.$$

By reason of symmetry there can be no difference of po-tential in the ring; but if the ring be cut, there immediately occurs at the points of separation a difference of potential representing the sum of the electromotive forces of the dif-ferent elements.   This case shows that an E. M. F. can ex-ist without difference of potential.

Hydrodynamics offers analogous examples: a circular

trough filled with water, in which a solid ring is revolved, would show a current due to the friction of the liquid against the ring; the motive force, being uniformly distributed, would not produce any difference of level between the various parts of the water in the trough.

188. Flux of Force Producing Induction.—It is important to determine the expression for the flux traversing an induced circuit.

In a general way the flux can be separated into two parts, the first due to the current itself which traverses the circuit, the second to the external field produced by currents or magnets.

In § 163 we defined, under the name of coefficient of self-induction of a circuit, the ratio of the flux traversing it to the current. This coefficient depends on the form of the circuit and on the medium in which it is placed. In fact, the flux of magnetic force generated is proportional to the permeability of the surrounding medium. If the circuit is completely surrounded with iron, or if it simply contains a core of iron, the flux of force has a very much higher value than if the circuit is simply surrounded by air; the permeability of iron for moderate inductions being very much greater than that of air.

If, then, we denote by $L_i i$ the flux of force produced by the current across its own circuit, the coefficient of self-induction L_i has a constant value only on condition that the circuit is invariable in form and placed in a feebly magnetic medium. Where the circuit is near very magnetic bodies, its coefficient L_i becomes variable with the current.

We cannot, therefore, specify the coefficient of self-induction of an electromagnet without specifying the current which traverses the field-coils, together with the former magnetic condition of the core.

As an application of the above, let us take an annular solenoid, in which the thickness of the section is negligible compared to the diameter, § 153.

Denoting by n the number of turns of wire per centimetre, measured along the circular axis of the solenoid, and by s its cross-section, the internal magnetic flux, for a current i, is expressed by

$$\mathcal{H}s = 4\pi n_1 is.$$

This flux traverses successively the n_1 turns of the solenoid; consequently the total flux across the solenoid is $\Phi = n \times \mathcal{H}s$, and the coefficient of self-induction has the value

$$L_s = \frac{\Phi}{i} = 4\pi n_1 ns.$$

Such a solenoid having 20 turns per unit of length and a section of 100 cm² would have a coefficient of self-induction per centimetre equal to $160,000\pi$ C. G. S. units or 0.503×10^{8} quadrants.

If there is an annular iron core of permeability μ in the solenoid, the coefficient becomes

$$L_s' = 4\pi n_1 n \mu s.$$

The permeability sometimes exceeds 3000, which explains why an electromagnet produces extra-currents very much greater than those of a solenoid without a core.

The preceding expression also represents the coefficient of self-induction of a straight electromagnet of great length, if we neglect the influence of the extremities.

When a coil is wound with wire doubled on itself, a generator of electric energy sets up in it helicoid currents of opposite directions whose resultant magnetic effect upon an interior core, as well as on the surrounding medium, is zero. It follows that such a solenoid has a negligible coefficient of self-induction. The same result is reached with a single wire if the successive layers are wound in the opposite directions, supposing that there are an even number of layers having each the same number of turns.

So, too, two straight conductors near together, or two wires twisted together, do not give rise to effects of lateral induction when they are traversed by currents equal and opposite in direction.

When the current passing in an electromagnet does not exceed the value which corresponds to the "elbow" in the core's magnetism curve, § 57, we may take for granted, in approximate calculations, that the permeability is constant and consequently the coefficient of self-induction also.

According to Lord Rayleigh this hypothesis always holds good, whatever be the magnetization of the core, when we are treating small variations of the magnetizing force, i.e. of the current.

To sum up, the total flux across an isolated circuit traversed by a current i is

$$\Phi = L_i i.$$

If the current varies, an E. M. F. of self-induction is set up, equal to

$$-\frac{d\Phi}{dt} = -\frac{d}{dt}(L_i i),$$

or simply

$$-L_i \frac{di}{dt},$$

if L_i is constant.

When the current increases, di is positive and the E. M. F. negative; whence an inverse extra-current. When the current is decreasing, the E. M. F. is positive and gives rise to a direct extra-current.

If the circuit, instead of being isolated, is near magnets or currents, to the flux due to the current itself is added a supplementary flux, comprising lines of force in the magnetic field produced by these external causes.

In the case of currents, we have defined (§ 142), under the name of *coefficient of mutual induction* of two circuits, the ratio of the flux traversing one of these circuits to the current in the other circuit. We must introduce the same condition here as in the case of self-induction; when the permeability of the medium is variable, L_m varies with the currents under consideration.

Suppose a circuit whose own flux is $L_s i$ and which is near another circuit whose coefficient of mutual induction and current are respectively L_m and i'. The total flux will be

$$\Phi = L_s i + L_m i'.$$

The E. M. F. of induction will in this case be expressed by

$$e = -\frac{d\Phi}{dt} = -\frac{d}{dt}(L_s i + L_m i').$$

If the two circuits are of invariable form, and if we suppose that the permeability of the medium surrounding them is constant, L_s and L_m are constant factors, and we have

$$e = -L_s \frac{di}{dt} - L_m \frac{di'}{dt}.$$

The secondary circuit may, moreover, be situated in a field produced by the earth and magnets. If we denote by $\mathcal{H}s + \Sigma(m\omega)$ the total flux set up by these causes across the circuit, we get the general expression

$$\Phi = L_s i + L_m i' + \mathcal{H}s + \Sigma(m\omega),$$

whence

$$e = -\frac{d}{dt}[L_s i + L_m i' + \mathcal{H}s + \Sigma(m\omega)].$$

NOTE.—Let us take up again for a moment the case of an annular bobbin, with an iron core closed upon itself and traversed by currents whose direction varies periodically, passing from $+ i$ to $- i$ and inversely. At every change of direction the magnetization of the core is reversed and the coefficient of self-induction for the bobbin takes the value

$$L_1 = 4 \Phi n n_1 \mu s.$$

But if the currents traversing the coils vary from o to a value i without assuming negative values, i.e. if the current is simply intermittent, the magnetic filaments formed in the core remain oriented by virtue of their coercive force, and the coefficient of self-induction of the bobbin has, for the currents following the first, sensibly the same value as if the core did not exist. Matters go on as if the core were a permanent magnet and the coefficient of self-induction simply equal to

$$L_2 = 4\pi n n_1 n s.$$

When the core is not continuous, no closed filaments are formed, so that the effect of the coercive force is considerably weakened.

189. Quantity of Induced Electricity.—Suppose the flux across a circuit varies from o to a quantity Φ. The induced current at any instant is

$$i = \frac{- \dfrac{d\Phi}{dt}}{r},$$

and the total quantity of induced electricity

$$q = \int_0^t i\, dt = \int_0^\Phi - \frac{d\Phi}{r} = - \frac{\Phi}{r}.$$

If the flux then returns to o, the quantity of electricity is

$$\int_{\bullet}^{\bullet} -\frac{d\Phi}{r} = \frac{\Phi}{r};$$

it is equal to the preceding, and is displaced in the opposite direction.

APPLICATIONS OF THE LAWS OF INDUCTION.

190. Movable Conductor in a Uniform Field.—Suppose that a conductor of length l, such as the axle of a railway car, is displaced horizontally with a velocity v. It cuts the terrestrial lines of force, and the E. M. F. of induction is, denoting by $\mathcal{H}$ the vertical component of the magnetic field,

$$e = -\frac{d\Phi}{dt} = -lv\mathcal{H}.$$

The direction of this E. M. F. is given by **Faraday's** rule (§ 186).

Put
$$l = 150 \text{ cm},$$

$$v = 1666 \text{ cm per second},$$

$$\mathcal{H} = 0.44.$$

We get $e = 150 \times 1666 \times 0.44$ C. G. S. units, or 0.0011 volt.

191. Faraday's Disc.—If Barlow's wheel is caused to rotate (§ 157), there arises an E. M. F. directed from the periphery to the centre or inversely, according to direction of rotation.

Keeping the notation used in the above-mentioned paragraph, and denoting by n the number of rotations per second, and by ω the angular velocity of the wheel,

$$\omega = 2\pi n;$$

the E. M. F. of induction is

$$e = \mathcal{H}sn = \frac{\mathcal{H}r^2\omega}{2}.$$

Faraday, to whom this experiment is due, thus discovered the simplest known induction-machine. We also are indebted to him for the following arrangement, which is the principle of the so-called unipolar machines.

Let us consider the apparatus shown in Fig. 75, and suppose that we leave out the cell, joining the trough directly to the middle of the magnet. If the conductor is now made to revolve, it becomes the seat of an E. M. F. of induction, whose value is, denoting by the ω angular velocity of the conductor, n the number of turns per second, and m the magnetic mass of the pole which serves as pivot,

$$e = 4\pi m \times n = 2m\omega.$$

Edlund has pointed out that, by leaving the conductor at rest and rotating the magnet, there is also an induced current set up. This can only be attributed to the development of an E. M. F. of induction in the magnet itself in consequence of its rotation in its own field.

192. Measurement of the Intensity of the Magnetic Field by the Quantity of Electricity Induced.—Suppose a uniform field of sufficient extent to allow a flat bobbin to turn upon itself in the field, its axis of rotation being along a diameter normal to the direction of the field. The ends of the wire in the coil are connected by sliding contacts to a ballistic galvanometer (§ 150).

Let R be the total resistance of the circuit, $\mathcal{H}$ the field-intensity. If the bobbin, containing n turns of surface a, is made to perform half a revolution, starting from a position with its plane normal to the direction of the field, the flux traversing it passes from $\mathcal{H}an$ to 0, then from 0 to $-\mathcal{H}an$. The total variation is $2\mathcal{H}an$, and the quantity of electricity induced (§ 171)

$$q = \frac{2\mathcal{H}an}{R}.$$

Again, denoting by α the swing of the galvanometer, by K a constant,

$$q = K\alpha;$$

whence

$$\mathcal{H} = \frac{K\alpha R}{2an}.$$

Weber's inclinometer, which serves to determine the intensity of the earth's field, is based upon this method.

If, by small equal advances, we slide a test-coil along a magnet which it exactly encloses, we will get in a ballistic galvanometer connected with the test-coil, swings proportional to the field-intensity in the various regions of the magnet. This is a convenient way to obtain the distribution-curve of magnetism in a magnet (§ 47).

193. Expression for the Work Absorbed in Magnetization. Loss due to Hysteresis.—Take the case of an annular electromagnet (§ 142) of sufficient diameter to allow the supposition that the internal field is uniform and expressed by

$$\mathcal{H} = 4\pi n_1 i,$$

n_1 being the number of turns per unit length.

The work done to magnetize the electromagnet is necessarily equal to the energy given back by it under the form of the extra-current. Now the energy of this latter is given by

$$W = \int_i eidt = \int_i^\bullet -\frac{d\Phi}{dt}\,idt = \int_i^0 -id\Phi.$$

But, denoting by s the cross-section of the magnet, l its mean length, and $\mathfrak{B}$ the induction traversing it, we have

$$\Phi = n_1 l s \mathfrak{B}.$$

Substituting for Φ this value, and for i, $\dfrac{\mathfrak{H}}{4\pi n_1}$, we get

$$W = -\frac{1}{4\pi} l s \int_{\mathfrak{H}}^{0} \mathfrak{H} d\mathfrak{B}.$$

The energy expended per unit of volume will be

$$-\frac{1}{4\pi} \int_{\mathfrak{H}}^{0} \mathfrak{H} d\mathfrak{B}.$$

Now $$\mathfrak{B} = \mathfrak{H} + 4\pi\mathfrak{J},$$
whence

$$-\frac{1}{4\pi} \int_{\mathfrak{H}}^{0} \mathfrak{H} d\mathfrak{B} = -\frac{1}{4\pi} \int_{\mathfrak{H}}^{0} \mathfrak{H} d\mathfrak{H} + \int_{\mathfrak{H}}^{0} \mathfrak{H} d\mathfrak{J}.$$

The integral

$$-\frac{1}{4\pi} \int_{\mathfrak{H}}^{0} \mathfrak{H} d\mathfrak{H} = \frac{\mathfrak{H}^2}{8\pi}$$

represents the energy restored by the solenoid, that is to say its intrinsic energy.

The integral

$$-\int_{\mathfrak{H}}^{0} \mathfrak{H} d\mathfrak{J}.$$

is therefore the energy restored by demagnetizing the core.

If the magnetizing force varies between values $\mathfrak{H}_1$, $\mathfrak{H}_2$, this last expression of energy will be

$$\int_{\mathfrak{H}_1}^{\mathfrak{H}_2} \mathfrak{H} d\mathfrak{J}.$$

It would be represented by the area comprised between the curve $\mathfrak{J} = f(\mathfrak{H})$, the axis of $\mathfrak{J}$ and two parallels to the

axis of abscissæ drawn through the points of the curve corresponding to $\mathcal{H}_1$ and $\mathcal{H}_2$.

Where the current successively assumes values $+i, -i,$
the energy restored by the solenoid equals the energy absorbed, for we have

$$-\frac{1}{4\pi}\int_{\mathcal{H}}^{\mathcal{H}} \mathcal{H}d\mathcal{H} = 0.$$

But the integral

$$-\int_{\mathcal{H}}^{\mathcal{H}} \mathcal{H}d\mathcal{B}$$

is not zero, since it represents the area comprised between
the curves ACA' and $A'C'A$ of Fig. 90.

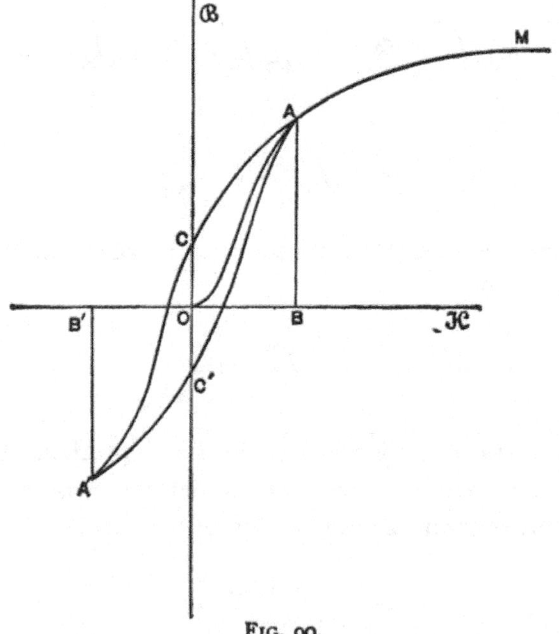

FIG. 90.

This integral expresses the loss per cycle and per cm³ of
the core.

The work expended, equal and of contrary sign to the work restored by the core, is represented by

$$\int_{\mathfrak{K}}^{'\mathfrak{K}} \mathfrak{K}d\mathfrak{I}.$$

194. Self-induction in a Circuit composed of Linear Conductors. Case of a Constant Electromotive Force. Time-constant.—When a circuit containing an element having constant E. M. F. is closed, the current does not instantaneously attain its normal value, especially if the circuit contains an electromagnet.

In the same way, when the circuit is broken the current does not stop suddenly, but is prolonged by the extra-current which appears in the spark at the point of interruption. Faraday showed the analogy existing between these phenomena and those caused by the inertia of fluids. A liquid current cannot be set up or cease suddenly in a pipe, and at the moment of stoppage we observe a sudden blow due to the momentum of the moving fluid; but there are profound divergences between these two cases.

While the blow given by a liquid is diminished by bends in the pipe, the extra-current is much more marked in a coil than in a straight wire of equal length. It will be seen further on that this difference is explained by placing the seat of the current's energy in the medium surrounding the conductors.

There exist, however, even in the expressions for the phenomena, analogies which it is useful to bring out.

Thus, when a fluid is urged to move in a pipe the force expended is utilized on the one hand to overcome the friction against the walls, on the other to increase the momentum of the movable mass. If the velocity v is small, the friction can be expressed by Av, when A is a constant.

The increase of momentum of the mass m is in unit time $m\dfrac{dv}{dt}$. The total force is therefore

$$F = Av + m\frac{dv}{dt}. \quad \ldots \ldots \quad (1)$$

Now take the case of a circuit with a resistance r, whose coefficient of self-induction L_s is constant (§ 198), and which includes a cell with an E. M. F. denoted by E.

At the moment of closing the circuit there is set up an E. M. F. of induction $-L_s\dfrac{di}{dt}$, so that the current is given by the equation

$$i = \frac{E - L_s\dfrac{di}{dt}}{r}.$$

From this we deduce

$$E = ri + L_s\frac{di}{dt}, \quad \ldots \ldots \quad (2)$$

an expression analogous to (1). ri represents, likewise, the portion of E. M. F. used to overcome the friction of the conductor and $L_s\dfrac{di}{dt} = e$ the portion used to increase the intrinsic energy of the circuit, for, multiplying e by i, we have

$$ei = L_s\frac{di}{dt}i = \frac{d}{dt}\left(\frac{Li^2}{2}\right).$$

It is taken for granted in equation (2) that the distribution of the current is uniform over the section of the conductor, and that the resistance of the latter to a variable current is the same as to a steady current. Further on we shall see that this is not the case for linear conductors, i.e. those of very small cross-section.

At the end of a time t the current will attain a value i

given by the integration of the differential equation (2). We have

$$\int_0^i \frac{di}{E - ir} = \int_0^t \frac{dt}{L},$$

whence

$$-\frac{1}{r} \log_e \frac{E - ir}{E} = \frac{t}{L},$$

and lastly

$$i = \frac{E}{r}\left(1 - e^{-\frac{rt}{L}}\right),$$

in which e represents the base of the Naperian logarithms.

Theoretically, the current does not reach its full value $\frac{E}{r}$ until an infinite time has elapsed, but as the value of $e^{-\frac{rt}{L}}$ decreases rapidly, this term becomes negligible, compared with unity, in a short time.

The ratio $\frac{L}{r}$, which is homogeneous with a time, is called the *time-constant* of the circuit; we will denote it by τ.

Suppose, for example, an annular solenoid of 100 cm^2 section, having 20 turns per cm, measured on the axis, and whose length when developed is 100 cm. If the resistance be 1 ohm or 10^9 C. G. S. units, the exponential factor assumes the value

$$e^{-\frac{10^9 t}{4\pi \times 20 \times 2000 \times 100}} = e^{-\frac{10^3 t}{16\pi}}.$$

It is seen that for a comparatively small value of t this term may be neglected.

The variation of the current in terms of the time is represented by a curve which rises rapidly from the origin, then tends towards an asymptote parallel to the axis of the times.   The first part of the curve, corresponding to small values of the time, can be replaced by a right line,

$$i = \frac{Et}{L_i}.$$

This equation shows that during the first instants the current depends, not on the resistance of the circuit, but on its self-induction.

The quantity of electricity which passes in the circuit during the variable period is

$$q = \int_0^t i\,dt = \int_0^t \frac{E}{r}\left(1 - e^{-\frac{t}{\tau}}\right) dt = \frac{E}{r}t - \frac{E}{r}\int_0^t e^{-\frac{t}{\tau}}\,dt$$

$$= \frac{E}{r}(t - \tau) + \frac{E}{r}\tau e^{-\frac{t}{\tau}}.$$

Beyond a certain value of t the second term becomes negligible, and we have simply

$$q = \frac{E}{r}(t - \tau).$$

$\frac{E}{r}t$ is the quantity of electricity which would have passed during the time t if the current had instantaneously assumed its permanent value.   The term $\frac{E}{r}\tau$ represents the quantity due to the inverse extra-current or make-induced current.

We come now to the break-induced current, and, to simplify the calculation, suppose that the resistance of the circuit be maintained constant by substituting for the cell a wire having the same resistance.

The break extra-current is given by the equation

$$i = \frac{-L'\frac{di}{dt}}{}.$$

Integrating between $\dfrac{E}{r}$ and i, we find

$$i = \frac{E}{r} e^{-\frac{t}{\tau}}.$$

It is easy to see that this current decreases rapidly.

The quantity of electricity due to the extra-current on breaking is the same as that of the extra-current on making the circuit.

The total quantity of electricity produced by the cell is, moreover, the same as if there were no induction-effects, for

$$Q = q + q' = \frac{E}{r}(t - \tau) + \frac{E}{r}\tau = \frac{E}{r}t.$$

We can find directly the expression for the quantity of induced electricity by applying the equation

$$q = \mp \frac{\Phi}{r}, \qquad (\S\ 189)$$

and observing that Φ is the product of the current, when fully established, into the coefficient of self-induction.

We see that the induction phenomena which occur in a circuit during the *variable period* of the current have the effect of causing an apparent increase in the resistance of the conductors. The total quantity of electricity put in motion is, however, the same whatever be the coefficient of self-induction, for the direct extra-current restores the quantity of electricity abstracted at the beginning of the current.

195. Work accomplished during the Variable Period. —By Joule's law, the work accomplished during the variable period on closing the circuit is expressed by

$$W_1 = \int_0^t i^2 r\, dt = r \int_0^t \frac{E^2}{r^2}\left(1 - e^{-\frac{t}{\tau}}\right)^2 dt$$

$$= \frac{E^2}{r}t - \frac{3}{2}\frac{E^2}{r}\tau + \frac{E^2}{r}\tau\left(2e^{-\frac{t}{\tau}} - \frac{1}{2}e^{-\frac{2t}{\tau}}\right).$$

If t is large enough, we have simply

$$W_1 = \frac{E^2}{r}\left(t - \frac{3}{2}\tau\right).$$

Now the cell produces during the variable period a total quantity of electricity

$$W = Eq = \frac{E^2}{r}(t - \tau).$$

This quantity exceeds the Joule effect by

$$\frac{E^2 L_s}{2r^2} = \frac{L_s I^2}{2},$$

denoting by I the current when fully established. This difference represents the intrinsic energy of the circuit traversed by the electric flux (§ 144).

We see, then, that when a current is set up in a circuit, the energy furnished by the source of electricity is composed of two parts, the one transformed directly into heat by the Joule effect, the other stored up in the potential state.

This intrinsic energy is in its turn transformed into heat during the extra-current on breaking the circuit, for there is then a Joule effect represented by

$$W_2 = \int_0^\infty i^2 r\, dt = \frac{E^2}{r}\int_0^\infty e^{-\frac{2t}{\tau}} dt = \frac{L_s I^2}{2}.$$

The potential energy of the current, according to modern ideas, resides in a special state of tension or movement of the magnetic field.

It corresponds, as we have seen, § 194, to the kinetic energy of a mass in motion.

Second Demonstration.—The expression for the intrinsic energy of a circuit, measured by the difference between the energy furnished by the source and that absorbed by the

Joule effect during the variable period, **can be deduced di-rectly** from the general law of induction

$$i = \frac{E - L_i \dfrac{di}{dt}}{r};$$

for we can write this in the form

$$\int_0^t E i\, dt - \int_0^t i^2 r\, dt = \int_0^I L_i i\, di = \frac{L I^2}{2}.$$

196. Application to the Case of Derived Currents.— Suppose two conductors joined in parallel, the self-induction of the first being L_i, that of the second having a negligible value; suppose further that the two branches have no mutual inductive influence on each other. When we send a current into the two conductors, whose resistances are r_1 and r_2, the division of electricity between them is influenced by the reaction of the self-induction in one of them. Let i_1 be the current in this one at any instant of the variable period, and i_2 the current in the other at the same instant.

The first one is the seat of an E. M. F. equal to $- L_i \dfrac{di_1}{dt}$, so that on applying Kirchhoff's second law to the closed circuit including r_1 and r_2, we get

$$i_1 r_1 - i_2 r_2 = - L_i \frac{di_1}{dt};$$

whence

$$r_1 \int i_1\, dt - r_2 \int i_2\, dt = - L_i \int di_1,$$

the integration being comprised between corresponding limits of time and current.

If the circuit of the cell is closed and then opened, the

initial and final currents are zero in the branch r_1; consequently the second member

$$- L_1 \int_0^\infty \mathrm{d}i_1 = 0$$

and

$$r_1 q_1 = r_2 q_2,$$

q_1 and q_2 denoting the quantities of electricity that have traversed the two branches.

These quantities are exactly the same as if there had been no induction effect in the two branches.

The same effect would occur if a condenser were discharged through the branches. The division would occur as if there were no self-induction currents in one of them.

It is taken for granted, in these deductions, that the conductors have a sufficiently small cross-section for the current to be uniformly spread over every part of it during the variable period.

197. Discharge of Condenser into a Galvanometer with Shunt.—The above calculations cannot be applied without reserve in the case of discharging a condenser into a galvanometer furnished with a shunt. If the galvanometer remains motionless during the whole period of discharge, the subdivision of the quantities of electricity takes place according to the above simple law. But it frequently happens that the needle commences to move before the end of the discharge; in which case it produces in the galvanometer-coil an induced current opposite in direction, by Lenz' law, to that which would itself have produced the movement. It follows that the total quantity of electricity which traverses the coils is diminished.

To calculate this diminution we may suppose, with M. L. Clark, that the flux of magnetic force produced by the needle across the coils is proportional to the sine of the angle described.

Denote by i_1, g, L_1 the current in the galvanometer, its resistance and coefficient of self-induction; by i_2, s, the current in the shunt and its resistance, and suppose it to be formed of a straight wire or bobbin with double winding. Kirchhoff's second law shows that

$$i_2 s - i_1 g = L_1 \frac{\mathrm{d}i_1}{\mathrm{d}t} + K \frac{\mathrm{d}\sin\alpha}{\mathrm{d}t},$$

whence

$$s \int_0^t i_2 \, \mathrm{d}t - g \int_0^t i_1 \, \mathrm{d}t = L_1 \int_0^0 \mathrm{d}i_1 + K \int_0^\delta \mathrm{d}\sin\alpha.$$

The current in the galvanometer is zero at the beginning and end of the discharge, whose duration is t, and the superior limit of α is the deviation δ of the needle.

Let

$$\int_0^t i_1 \, \mathrm{d}t = q_1, \qquad \int_0^t i_2 \, \mathrm{d}t = q_2,$$

we will have

$$s q_2 - g q_1 = K \sin \delta.$$

But by the theory of the ballistic galvanometer (§ 150), if the arc of swing is small enough

$$\sin \delta = 2 \sin \frac{\delta}{2} = a q_1,$$

a being a constant.

Consequently

$$s q_2 - g q_1 = a K q_1.$$

Denoting by Q the total discharge equal to $q_1 + q_2$,

$$s(Q - q_1) - g q_1 - a K q_1 = 0,$$

whence

$$Q = \frac{q_1 (g + s + a K)}{s}.$$

198. Self-induction in a Circuit of Linear Conductors where there is a Periodic or Undulatory Electromotive Force.—From a practical point of view a very important case of induction occurs when one or more spirals of wire are rotated in a magnetic field.

For simplicity's sake, suppose a coil *abgd* turns about the axis *bd* (Fig. 85) in a direction opposite to the movement of the hands of a watch, in a uniform field whose direction is normal to the plane of the coil in its present position.　During a semi-revolution the part *bgd* will cut the lines of force in one direction, the part *dab* in the opposite.　The E. M. Fs. produced will be added together and produce an induction current in the coil.　During the next half-revolution the direction of displacement of the two halves of the coil will be reversed and consequently the induced current will change its direction in the coil.

In one complete revolution, the current is therefore reversed twice in the coil.　It is easy to see that the E. M. F. is maximum at the moment when the lines of force are cut normally, i.e., when the plane of the coil is parallel to the direction of the field.

Let us apply the general law of induction to this case.

Let $\mathcal{H}$ be the field-intensity, S the surface of the coil, $a = \dfrac{d\alpha}{dt}$ its angular velocity, supposed to be constant.　The angle described in a time t will be $\alpha = at$.

When the plane of the coil makes an angle α with the direction normal to the field, the flux traversing the coil is

$$S\mathcal{H} \cos \alpha = \Phi.$$

The E. M. F. due to the field is

$$E = -\frac{d\Phi}{dt} = S\mathcal{H} \sin \alpha \frac{d\alpha}{dt} = S\mathcal{H}a \sin \alpha = S\mathcal{H} \sin at.$$

Let T be the *period*, that is, the duration of one complete revolution, corresponding to an angle 2π, and n the *frequency* or *number of periods* per second. The number of

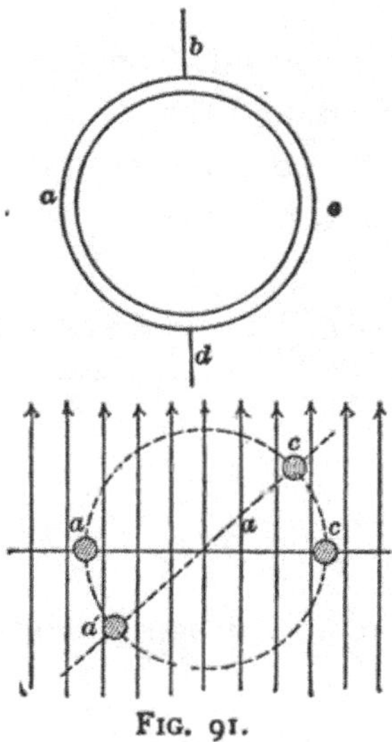

FIG. 91.

alternations is double the number of periods. We can write indifferently

$$E = \mathfrak{SK}a \sin at = \mathfrak{SK}a \sin\frac{2\pi}{T}t = \mathfrak{SK}a \sin 2\pi nt.$$

For the values

$$t = \frac{T}{4}, \ \frac{3T}{4}, \ \frac{5T}{4},$$

we have

$$\sin\frac{2\pi}{T}t = +1, \ -1, +1.$$

The E. M. F. passes successively through maxima

$$E_0 = \mathfrak{SK}a$$

and through minima

$$-E_0 = -\mathfrak{SK}a.$$

The expression representing the variation of the E. M. F. as a function of the time can therefore be put in the form

$$E = E_{\text{o}} \sin at. \qquad \ldots \ldots \quad (1)$$

The current in the coil results from the combination of the E. M. F. due to the field, and the E. M. F. produced by the self-induction L_{s} of the coil, whose resistance is r; consequently, if L_{s} is constant, we have

$$i = \frac{E - L_{\text{s}}\dfrac{di}{dt}}{r} = \frac{E_{\text{o}} \sin at - L_{\text{s}}\dfrac{di}{dt}}{r}, \qquad \ldots \quad (2)$$

whence

$$di + \frac{r}{L_{\text{s}}}i dt = \frac{E_{\text{o}}}{L_{\text{s}}} \sin at dt, \qquad \ldots \ldots \quad (3)$$

If we put $i = uv$, u and v being arbitrary variables, we get

$$u\!\left(dv + \frac{r}{L_{\text{s}}}v dt\right) + v du = \frac{E_{\text{o}}}{L_{\text{s}}} \sin at dt. \qquad \ldots \quad (4)$$

Now make

$$dv + \frac{r}{L_{\text{s}}}v dt = 0,$$

and for simplicity put $\dfrac{r}{L_{\text{s}}} = b$, when we get

$$\log_e v = -\int b dt + \log_e K,$$

$\log_e K$ being a constant of integration; whence

$$v = Ke^{-\int b dt}. \qquad \ldots \ldots \ldots \quad (5)$$

Now equation (4) reduced to

$$v du = \frac{E_{\text{o}}}{L_{\text{s}}} \sin at dt$$

gives

$$du = \frac{1}{K} e^{\int b dt} \frac{E_0}{L_s} \sin at \, dt;$$

whence

$$u = K' + \frac{1}{K} \int e^{\int b dt} \frac{E_0}{L_s} \sin at \, dt;$$

and putting

$$KK' = D,$$

$$i = uv = e^{-bt} \left\{ \int e^{bt} \frac{E_0}{L_s} \sin at \, dt + D \right\}$$

$$= e^{-bt} \frac{E_0}{L_s} \int e^{bt} \sin at \, dt + D e^{-bt}.$$

Integrating by parts, we get

$$\int e^{bt} \sin at \, dt = \frac{e^{bt}}{a^2 + b^2} (b \sin at - a \cos at),$$

whence

$$i = \frac{E_0}{L_s(a^2 + b^2)} (b \sin at - a \cos at) + D e^{-bt}. \quad . \quad (6)$$

We can substitute for the difference

$$\frac{b}{\sqrt{a^2 + b^2}} \sin at - \frac{a}{\sqrt{a^2 + b^2}} \cos at$$

by sin $(at - \phi)$, the value of ϕ being determined by the condition that the equality

$$\frac{b}{\sqrt{a^2 + b^2}} \sin at - \frac{a}{\sqrt{a^2 + b^2}} \cos at = \sin (at - \phi)$$

be verified for all the values of t.

Now for

$$t = 0 \text{ we get } \quad \sin \phi = \frac{a}{\sqrt{a^2 + b^2}},$$

for

$$t = \frac{\pi}{2a}, \qquad \cos \phi = \frac{b}{\sqrt{a^2 + b^2}};$$

therefore

$$\phi = \tan^{-1} \frac{a}{b} = \tan^{-1} \frac{aL_s}{r}.$$

The term De^{-bt} of equation (6) expresses the increase of the current during the first moments the E. M. F. is acting. After a few instants this term becomes negligible, and the current is given by

$$i = \frac{E_o}{L_s(a^2 + b^2)} \sin (at - \phi). \quad . \quad . \quad . \quad (7)$$

Substituting for a and b their values, we get finally

$$i = \frac{E_o}{\sqrt{r^2 + \dfrac{4\pi^2 L_s^2}{T^2}}} \sin \left(\frac{2\pi t}{T} = \phi\right), \quad . \quad . \quad . \quad (8)$$

on the condition that

$$\phi = \tan^{-1} \frac{2\pi L_s}{rT}. \quad . \quad . \quad . \quad . \quad . \quad (9)$$

We could also give equation (8) the form

$$i = \frac{E_o}{r\sqrt{1 + \tan^2 \phi}} \sin \left(\frac{2\pi t}{T} - \phi\right)$$

$$= \frac{E_o \cos \phi}{r} \sin \left(\frac{2\pi t}{T} - \phi\right). \quad . \quad . \quad . \quad (10)$$

Equation (10) shows that the maximum current is

$$\frac{E_o \cos \phi}{r} = i.$$

It is observed that the E. M. F. of self-induction reduces the maximum current, which would be $\dfrac{E_0}{r}$ if L_1 were zero.

$E_0 \cos \phi$ represents the maximum effective E. M. F. resulting from the combination of the electromotive force E_0 and the reaction of the self-induction. The subtractive term ϕ, which is not found in equation (1) of E. M. F., shows that there is a retardation of phase between the maximum values of the current and the E. M. F. due to the field. This retardation is $\dfrac{T}{2\pi}\phi$ in duration.

Remarks.—I. These results are applicable to a coil composed of a number of turns, and, for approximate calculations, to a coil with an iron core, provided the magnetization be sufficiently feeble to allow us to take the permeability as constant, without much error.

II. We denote by the name of *apparent resistance of a circuit* the radical

$$\sqrt{r^2 + \frac{4\pi^2 L^2}{T^2}},$$

by which the E. M. F. must be divided in order to find the current : Heaviside has called this the *impedance* of the circuit. It will be noticed that the radical is homogeneous with a resistance and can be expressed in ohms. The term $\dfrac{2\pi L_1}{T}$ is called the *reactance*.

III. Denoting by τ the time-constant of the circuit equal to $\dfrac{L_1}{r}$, the result can also be put in the form

$$i = \frac{E_0}{r\sqrt{1 + a^2\tau^2}} \sin (at - \phi),$$

$$\phi = \tan^{-1} a\tau.$$

These formulæ show that the apparent resistance (impedance) and the retardation of phase depend essentially on the time-constant. A large self-inductance may only produce a minimum apparent increase of resistance, if the resistance is already considerable in itself.

199. Graphic Representations.—The axis Oy representing the direction of a uniform field which develops induction-currents in a coil turning in the direction of the arrow, let us represent by the right line OM, which makes an angle $\alpha = \dfrac{2\pi t}{T}$ with Ox, the maximum electromotive force $E_{\circ}$. The E. M. F. due to the field is represented at the instant t by the projection $oa = E_{\circ} \sin \alpha$.

The maximum effective electromotive force, $E_{\circ} \cos \phi$, is represented by ON, the projection of OM on a straight line making an angle ϕ with the latter.

Now as the effective E. M. F. is the resultant of $E_{\circ}$ and

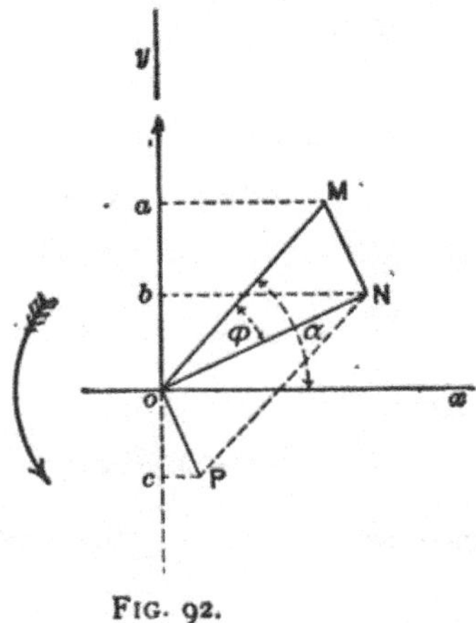

FIG. 92.

the E. M. F. of self-induction, the maximum of this last will be shown by a right line $OP = E_{\circ} \sin \phi$, which completes the parallelogram $OM . NP$. The projections of OM, ON

and OP on Oy represent the values of the different E. M. Fs. at the moment of rotation when the coil makes an angle α with Ox.

The effective E. M. F. at this moment is

$$Ob = ON \sin (\alpha - \phi) = E_0 \cos \phi \sin (\alpha - \phi);$$

it is less than the E. M. F., Oa, due to the field, by a quantity $ab = Oc$ which measures the reaction of self-inductance.

But when the coil OM has passed the axis Oy by an angle ϕ, OP comes above the axis of x and the projection of the resultant, ON, is greater than the projection of OM, for the action of the self-induction is now added to the E. M. F. due to the field.

On rotating the parallelogram $OMNP$ about the point O, the projections of OM, ON, and OP on Oy show, at each instant, the relative values of the various E.M.Fs. in action.

The current in the coil is given at any given instant by the ratio of the length of Ob to the resistance r of the coil.

We can also show the variations of the E. M. Fs. in terms of the time, by drawing the curves represented by the equations

$$E = E_0 \sin \frac{2\pi t}{T} = E_0 \sin at,$$

$$e = - L \frac{di}{dt} = - E_0 \sin \phi \cos (at - \phi),$$

$$E' = E_0 \cos \phi \sin (at - \phi).$$

We will then get three sinusoidal curves like those in Fig. 93; the sine-curve E represents the E. M. F. due to the field; e is the E.M.F. of telf-induction and E' the effective or resultant E. M. F. The ordinates of E' are the differences between those of E and e.

The current at any moment is found by dividing the ordinates of the curve E' by the resistance of the circuit ; the phase of the current, moreover, coincides with that of E'.

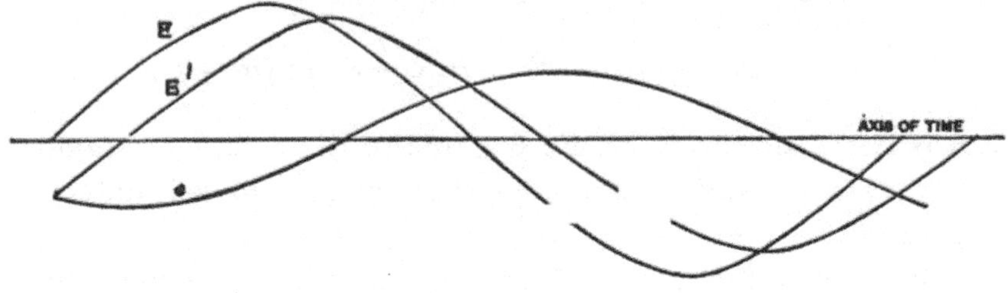

FIG. 93.

The retardation of the current-phase behind that of the E. M. F. due to the field is

$$\phi \frac{T}{2\pi} = \theta,$$

where

$$\phi = \tan^{-1}\frac{aL_s}{r}.$$

The maximum possible retardation corresponds to

$$\frac{aL_s}{r} = \infty .$$

We have then

$$\phi = \frac{\pi}{2} \quad \text{and} \quad \theta = \frac{T}{4}.$$

It will be observed that the curve e is a quarter phase, or $\frac{T}{4}$, behind the curve E'. Consequently if the latter is itself $\frac{T}{4}$ behind E, the positive waves in E will be exactly over the negative waves in e. The effective E. M. F. ought then to be zero (and also the current) for

$$E_0 \cos \phi = 0.$$

200. Mean Current and Effective Current. Measurement by Dynamometer.—The quantity of electricity which passes through the circuit in a half-period is independent of the lag; it is expressed by

$$q = \int_0^{\frac{\pi}{a}} i\, dt = \frac{E_0}{\sqrt{r^2 + a^2 L_s^2}} \int_0^{\frac{\pi}{a}} \sin at\, dt = \frac{2E_0}{a\sqrt{r^2 + a^2 L_s^2}}$$

The mean current, during a half-period, is

$$i_m = \frac{q}{\dfrac{\pi}{a}} = \frac{2}{\pi} \times \frac{E_0}{\sqrt{r^2 + a^2 L_s^2}} \frac{2}{\pi} I,$$

I being the maximum current.

The *mean current-strength*, which may be represented by

$$\frac{1}{\frac{1}{2}T} \int_0^{\frac{1}{2}T} i\, dt,$$

is therefore equal, in the case under consideration, to the product of the maximum current by the factor $\dfrac{2}{\pi}$.

It is to be observed that the needle of a galvanometer gives no deviation when the coil of the instrument is traversed by an alternating current of short period, for it then receives equal and contrary impulses; but it is possible to use the electrodynamometer, § 152, whose readings are proportional to the square of the current.

With this instrument we get a deviation proportional to the mean of the squares of the current.

This mean is expressed by

$$\frac{1}{T} \int_0^T i^2\, dt,$$

and its square root is called the *effective current*, the luminous effects of the current in electric lamps, for example,
depending upon it.

The *effective E. M. F.* is likewise defined by the square
root of the mean square of the E. M. F.

The *impedance* of the circuit is the factor by which the
effective current must be multiplied to obtain the effective
E. M. F.　Sometimes the name of *inductive resistance* is
given to a resistance having a coefficient of self-induction.

In the present case, the mean square of the current is

$$\frac{1}{\frac{1}{2}T}\int_0^{\frac{1}{2}T} i^2 dt = \frac{\int_0^{\frac{\pi}{a}} i^2 dt}{\frac{\pi}{a}}$$

$$= \frac{\frac{E_0^2}{r^2 + a^2 L_e^2}\int_0^{\frac{\pi}{a}} \sin^2 at\, dt}{\frac{\pi}{a}} = \frac{E_0}{r^2 + a^2 L_e^2} \times \frac{1}{2}.$$

It will be observed that the square root of the mean
square is different from the mean current,

$$\frac{i_m}{\sqrt{(i^2)_m}} = \frac{2\sqrt{2}}{\pi} = 0.9;$$

consequently

$$i_m = 0.9\sqrt{(i^2)_m};$$

that is to say, the readings by the electrodynamometer must
be reduced by a tenth to get the mean current.

If we denote by E_0 and I the E. M. F. and the maximum
current, it is easy to express in terms of these quantities
both the mean E. M. F. and current and the effective
E. M. F. and current.

We have then :

Mean E. M. F. $\qquad e_m = \dfrac{2}{\pi} E_0.$

Mean current $\qquad i_m = \dfrac{2}{\pi} I.$

Effective E. M. F. $\sqrt{\overline{(e^2)}_m} = E_{\text{eff}} = \dfrac{E_0}{\sqrt{2}}.$

Effective current $\sqrt{\overline{(i^2)}_m} = I_{\text{eff}} = \dfrac{i}{\sqrt{2}} = \dfrac{E_{\text{eff}}}{\sqrt{r^2 + \dfrac{4\pi^2 L_s^2}{T^2}}}.$

Impedance $\qquad \sqrt{r^2 + \dfrac{4\pi^2 L_s^2}{T^2}}.$

It must not be forgotten that these various relations do not hold except in the case where the periodic E. M. F. is a simple sine-function of the time and where the self-inductance L_s is constant ; which makes it necessary that the permeability of the surrounding medium be invariable.

If the periodic function were more complex, it might, by Fourier's theorem, be represented by a sum of sinusoids, but the above coefficients of reduction would be changed.

The *mean heat* developed by the current in one second is

$$\frac{1}{T} \int_0^T i^2 r \, dt = r \times \frac{1}{T} \int_0^T i^2 dt ;$$

that is, the product of the real (or " ohmic ") resistance of the circuit by the square of the effective current.   We have

$$P_m = \frac{1}{\dfrac{\pi}{a}} \int_0^{\frac{\pi}{a}} i^2 r \, dt = \frac{1}{2} \frac{E_0^2 r}{r^2 + a^2 L_s^2}$$

$$= \frac{1}{2} E_0 I \frac{r}{\sqrt{r^2 + a^2 L_s^2}} = \frac{1}{2} E_0 i \cos \phi,$$

whence

$$P_m = \sqrt{(e^2)_m}\,\sqrt{(i^2)_m}\,\cos\phi = E_{eff}I_{eff}\cos\phi.$$

We see that the heat developed in a circuit where a periodic E. M. F. is acting varies with the lag of the current-phase behind the E. M. F., and consequently with the self-induction of the circuit. The calorific power is zero for $\phi = \dfrac{\pi}{2}$, i.e., for a lag of $\frac{1}{4}$ of a period.

In the expression

$$p_m = \frac{1}{2}\frac{E_0^2}{r + \dfrac{a^2 L_s^2}{r}}$$

if L_s is constant, the denominator is minimum for

$$r = \frac{a^2 L_s^2}{r};$$

consequently the value of the resistance of the circuit which renders the mean calorific power a maximum is

$$r = aL_s.$$

As then

$$\tan\phi = \frac{aL_s}{r} = 1,$$

we have

$$\phi = \frac{\pi}{4},$$

and the lag corresponding to the maximum power is equal to $\frac{1}{8}$ period. This power is expressed by

$$p = \frac{E_0^2}{4r}.$$

˙201˙ Mutual Induction of Two Circuits.—Suppose that two circuits of invariable form, having resistances r and r' and a coefficient of mutual induction equal to L_m (§ 143), approach each other so slowly that the currents i, i', which traverse them, can be considered as constant.

For an elementary displacement, denoting by E, E' the electromotive forces of the sources producing the currents, we have

$$i = \frac{E - i' \dfrac{dL_m}{dt}}{r}, \quad i' = \frac{E' - i \dfrac{dL_m}{dt}}{r'}.$$

From these equations we get

$$(Ei + E'i')dt - (i^2 r + i'^2 r')dt = 2ii'dL_m.$$

$(Ei + E'i')dt$ expresses the energy furnished by the generators during a time dt; $(i^2 r + i'^2 r')dt$ is the portion of this energy which is transformed into heat in the conductors.

$ii'dL_m$ is the work of the electrodynamic forces. As the excess of the energy expended over that transformed into heat is double this work, we conclude that a portion equal to $ii'dL_m$ is stored up in the system in the state of potential, or intrinsic, energy.

It will be remembered, in fact (§ 143), that the mutual energy of two circuits is expressed by $-ii'L_m$; its variation is therefore, of course, $ii'dL_m$.

202. Mutual Induction of Two Fixed Circuits.—Two circuits, invariable in form and position, have resistances r and r', coefficients of self-inductance L, and L', and a coefficient of mutual inductance L_m (§ 143).

If cells having electromotive forces E and E' are in the circuits, the currents at any given instant of the variable period will be

$$i = \frac{E - \dfrac{\mathrm{d}}{\mathrm{d}t}(L_m i' + L_s \dot{i})}{r}, \qquad \cdots \cdots \quad (1)$$

$$i' = \frac{E' - \dfrac{\mathrm{d}}{\mathrm{d}t}(L_m i + L_s' i')}{r'}. \qquad \cdots \cdots \quad (2)$$

As the circuits are fixed, L_m, L_s and L_s' are constants, so that on multiplying (1) and (2) respectively by $i\,\mathrm{d}t$ and $i'\mathrm{d}t$ and then adding, we get

$$(Ei + E'i')\mathrm{d}t - (i^2 r + i'^2 r')\mathrm{d}t$$
$$= L_s i\,\mathrm{d}i + L_s' i'\mathrm{d}i' + L_m(i\,\mathrm{d}i' + i'\mathrm{d}i).$$

We see that, in this case, the excess of energy furnished by the cells over the energy transformed into heat is

$$L_s i\,\mathrm{d}i + L_s' i'\mathrm{d}i' + L_m(i\,\mathrm{d}i' + i'\mathrm{d}i).$$

This expression is the exact differential of

$$\frac{L_s i^2}{2} + \frac{L_s' i'^2}{2} + L_m i i',$$

which represents the potential energy of the circuits when the currents have values i, i'.

The first two terms represent the intrinsic energy of each circuit, and the third is their mutual energy.

203. Quantity of Induced Electricity.—Let us take the case where $E' = 0$; the current of the second circuit is then alone the cause of the mutual induction, and equation (2) gives by integration

$$\int_0^t r' i'\,\mathrm{d}t = - L_m \int_0^I \mathrm{d}i - L_s' \int_0 \mathrm{d}i',$$

for the induced current is zero at the beginning and end of the variable period of the inducing current.

Consequently

$$\int_{0}^{\infty} i' \,\mathrm{d}t = q_1' = -\frac{L_m}{r'}I = -\frac{L_m E}{rr'}.$$

When the inducing circuit is broken, we have likewise

$$q_2' = \int_{0}^{\infty} i' \,\mathrm{d}t = -\frac{L_m}{r'}\int_{I}^{0}\mathrm{d}i - \frac{L_s'}{r'}\int_{0}^{0}\mathrm{d}i' = \frac{L_m I}{r'} = \frac{L_m E}{rr'}.$$

The quantities of electricity induced are equal and of opposite sign in the two cases.

204. Expression for Mutual Inductance.—We have shown (§ 163) that the sum of the magnetic fluxes which traverse the coils of a cored annular bobbin, divided by the current in the coils, is expressed by

$$L_s = 4\pi n_1 n \mu s = 4\pi n_1^2 \mu l s,$$

when the diameter of the ring is very great compared to its thickness. If we suppose that the first bobbin is entirely covered by a second, having n_1' turns per unit of length along the axis, and n' total turns, the coefficient of induction of this second bobbin is

$$L_s' = 4\pi n_1' n' \mu s = 4\pi n_1'^2 \mu l s.$$

Now the coefficient of mutual induction or the mutual inductance is the ratio of the flux produced by one of the bobbins across the coils of the other, to the current which traverses the first.

$$L_m = 4\pi n_1 \mu s \times n' = 4\pi n_1' \mu s \times n = 4\pi n_1 n_1' \mu s l.$$

We see therefore that we then have the relation

$$L_m = \sqrt{L_s L_s'}.$$

This simple expression is applicable whenever the lines of force generated by one of the bobbins traverse all the turns of the other. It is the maximum value, therefore, of the mutual induction of the two circuits. This condition is realized in the middle region of two concentric solenoids of great length and with a rectilinear axis.*

205. Induction in Metallic Masses.—In what has pre ceded, we have had particularly in view the phenomena of induction developed in linear circuits ; but it is evident that induction takes place in metallic masses of any shape when they cut lines of force in a magnetic field.

Faraday's disc (§ 191) shows the development of induced currents in the case of a solid disc turning between the poles of an electromagnet. The determination of the lines of electric flux or current in such a case presents great complexity.

To resolve the problem, we put against the disc copper points connected with a galvanometer, proceeding as shown in § 168. In this way we get series of points at the same potential, which enables us to draw equipotential lines. The lines of flux are perpendicular to these.

If Faraday's disc is made to revolve without connecting the sliding contacts by a conductor, the lines of current be· come closed on themselves in the disc, producing curves which envelop each other without intersecting, and which are divided into two separate groups by a vertical plane passing through the axis of rotation.

By Lenz' law the direction of these currents is such that they tend to oppose the movement of the disc ; consequently the currents which approach the poles are in the opposite di- rection to the currents of a solenoid equivalent to the induc-

---

* See Mascart and Joubert, *Leçons sur l'Electricité et le Magnetisme* for the working out of calculations on inductances of solenoids.

ing magnet. The currents which move away from the poles are in the same direction as would be the solenoidal ones.

206. Foucault Currents. — This consequence of Lenz' law is shown in various ways; the following experiment is due to Foucault. A rapid movement of rotation is given to a disc embraced by the pole-pieces of a powerful electro-magnet. At the moment when a current is passed through the latter, the mechanical resistance caused by the induced currents causes the stoppage of the disc; if the motion is continued by a sufficient expenditure of motive power, the disc grows hot in consequence of the Joule effect.

The induced currents are very greatly reduced, and consequently the resistance to rotation and the heating, by dividing the disc by slits normal to the direction of the E. M. Fs. of induction and thus cutting the lines of flux. In the present case these divisions would be circles concentric with the disc, and this latter would have to be made of rings of increasing diameter, separated by an insulating material.

The currents induced in metallic masses are usually comprised under the name of *Foucault* or *eddy currents*.

The mechanical resistance caused by the induced currents in these masses is utilized to deaden the movement of galvanometer-needles.

If, for example, we surround a magnet, movable around an axis of suspension, with a mass of copper in which is hollowed a cavity sufficient to permit the swings of the magnet, the latter will develop in the copper induced currents which oppose its movement and cause its stoppage.

When a conductive rod which forms part of a circuit is displaced across the lines of force of a field presenting variations of intensity at different points, besides the E. M. F. of induction observed in the circuit, Foucault currents are generated in the substance of the conductor. In fact the elementary filaments which constitute the conductor cut at the same instant different numbers of lines of force. These subsidiary currents heat the rod without doing useful work in the circuit.

207. Cores of Electromagnets Traversed by Variable Currents. Calculation of the Power Lost in Foucault Currents.—Foucault currents tend to be set up in the core of an electromagnet whose coils are traversed by a periodic current. To diminish these currents, which heat the iron uselessly, the core is made of thin plates insulated from each other by varnish, or varnished or paraffined paper, and put together in such a way that their surfaces of separation are parallel to the axis of the coils and consequently cut the directions of the E. M. Fs. of induction. The bolts used to fasten the plates together should be insulated by tubes of vulcanized fibre and washers of the same put under the heads of the bolts and the nuts. It is a good plan to wrap the whole core round with varnished cloth, so that the edges of the plates may not pierce the insulation of the wires rolled on the iron.

A core is sometimes made of a bundle of varnished iron wires, but it is to be remarked that the space lost by the interstices between the wires is much greater than in the case of a laminated core. Moreover this arrangement is only applicable to small, straight electromagnets. As the wires do not touch except along single lines, the insulation need not be so careful as that of the plates. The division of the core parallel to its axis enables the heating by Foucault currents to be avoided, but not that which results from hysteresis (§ 61). Under the action of the periodic current of the coils, the core is, practically, subjected to successive magnetizations in opposite directions which cause a loss of energy in proportion to the variations of the magnetizing force and the coercive force of the core.

It is possible to calculate the loss occasioned by Foucault currents in an iron plate or wire taken in the core of an electromagnet traversed by periodical currents.

Take the case of a cylindrical wire of length l and radius R, traversed longitudinally by a variable flux which tends to develop eddy currents parallel to the edge of a right section of the wire. Let us take inside the wire a concentric tube infinitely thin, of radius r and thickness dr. Every variation of the flux along the wire sets up in the tube an E. M. F. of induction expressed by

$$e = -\frac{d\Phi}{dt} = -\pi r^2 \frac{d\mathfrak{B}}{dt}.$$

The resistance opposed to the current by the tube is

$$\frac{2\pi r \rho}{l\,dr},$$

ρ being the specific resistance of the metal.

The power set free in the form of heat in the tube is the ratio of the square of the E. M. F. to the resistance, or

$$dp = \frac{\left(-\pi r^2 \dfrac{d\mathfrak{B}}{dt}\right)^2}{\dfrac{2\pi r \rho}{l\,dr}} = \pi r^3 \frac{l}{2\rho}\left(\frac{d\mathfrak{B}}{dt}\right)^2 dr.$$

The total loss in the wire is, consequently,

$$p = \int_0^R \pi r^3 \frac{l}{2\rho}\left(\frac{d\mathfrak{B}}{dt}\right)^2 dr = \frac{\pi R^4 l}{8\rho}\left(\frac{d\mathfrak{B}}{dt}\right)^2.$$

The loss of power per cm.^{3} is

$$\frac{p}{\pi R^2 l} = \frac{R^2}{8\rho}\left(\frac{d\mathfrak{B}}{dt}\right)^2.$$

In the case of a plate of length l, thickness n, and breadth m, we would have found, supposing the lines of current parallel to the edges of the plate (which is only approximately exact),

$$\frac{p}{lmn} = \frac{n^2}{6\rho}\left(\frac{d\mathfrak{B}}{dt}\right)^2.$$

In each case we observe that the loss diminishes rapidly with the thickness of the elements of the core.

From the researches of Messrs. J. Thomson and Ewing [*]
it appears that the Foucault currents, produced in the elements of the core of an electromagnet traversed by alternating currents, exercise a magnetizing effect on the molecules of iron inside the plates or wires composing the core, which effect is the opposite of that exercised by the current in the field-coils. It follows that in order to obtain a given total magnetic flux, the magnetic induction in the outer layers of the wires or plates must be forced up, which causes an increase in the loss by hysteresis, since this latter increases more rapidly than the maximum induction to which the iron is subjected.

208. Self-induction in the Mass of a Cylindrical Conductor. Expression for the Coefficient of Self-induction in such a Conductor.—When a variable current flows in a cylindrical conductor, the current-density is not constant over the whole section of the conductor; it is greater at the periphery than at the centre. To account for this fact we can mentally divide the total current into an infinite number of elementary parallel currents, capable of reacting upon each other. The current which passes in one filament tends to induce inverse currents in the neighboring filaments. These mutual reactions are all the stronger since

---

[*] *Electrician*, April 22, 1892.

the filaments are crowded together. A reduction of the current therefore follows, which is a maximum towards the middle of the conductor's section, and minimum at the periphery. The conductor presents for variable currents a resistance which is higher than that with constant currents.

The induction in the mass of the conductor is further increased in the case of a magnetic wire, such as an iron one, in consequence of the circular magnetization which the metal then assumes, and which is due to the circular lines of force generated by the filaments of the current inside (§ 135). As was shown, however, in § 188, this effect does not appear except in the case of alternating currents; with intermittent currents, constant in direction, the molecules of the iron kept their orientation in the form of closed chains and do not produce any induction effect; with alternating currents it is preferable to use conductors of some non-magnetic metal.

When the period of variation is excessively short, as is the case in discharging a condenser, it may be conceived that the mutual reactions carry the whole current towards the external layers of the wire. In proportion as the period of the current decreases, we are thus led to separating the elementary filaments more and more and to giving the conductors the greatest surface possible. Metallic ribbons, tubes, and cords are then preferable to conductors of circular section.

To arrive at the value of the coefficient of self-induction of a cylindrical conductor, let us first find that of a circuit formed of two conductors G, G', parallel and long enough to be considered as indefinite. Such would be the case with two neighboring telegraph-wires. Let i be the current traversing the circuit, r the radius of the conductors, and d the distance between their axes.

Let us determine the flux of force produced by the conductors in the space limited by their axes and by two planes normal to them, 1 cm. apart.   Each of the conductors evidently produces half the flux.   Denote by μ the permeability of the surrounding medium, and by μ' that of the wires.

The conductor G produces in an external point, situated at a distance a from its axis, a field whose intensity is the same as if the current were condensed along the axis, or $\dfrac{2i}{a}$ (§ 146).   The corresponding magnetic induction is consequently $\dfrac{2\mu i}{a}$.   If we imagine in the above-defined space a section parallel to G, of thickness da, the flux across this elementary surface is $\dfrac{2\mu i \mathrm{d}a}{a}$.   The total flux, due to G, which cuts the surface comprised between the edge of G and the axis of G' is therefore per unit length

$$\int_r^{d} \frac{2\mu i \mathrm{d}a}{a} = 2\mu i \log_e \frac{d}{r}.$$

In the part of the space under consideration occupied by G, the field has a different expression.   In a point taken in the interior of G at a distance b from the axis, the field is the same as that which would be produced by a current condensed along the axis and bearing the same proportion to the total current as a cross-section of radius b bears to the total cross-section of the conductor.

We will then have for the field-intensity

$$\frac{2i}{b} \times \frac{\pi b^2}{\pi r^2} = \frac{2ib}{r^2},$$

and for the value of the magnetic induction at the point b

$$\frac{2\mu' ib}{r^2}.$$

The flux traversing half the longitudinal section of G, and which traverses the mass of the conductor, will be, per unit length along the axis,

$$\int_0^r \frac{2\mu' ib\, db}{r^2} = \mu' i.$$

The total flux due to G is therefore

$$i\left(2\mu \log_e \frac{d}{r} + \mu'\right).$$

As G' furnishes an identical flux across the space under consideration, we will get altogether

$$2i\left(2\mu \log_e \frac{d}{r} + \mu'\right).$$

By definition, the self-inductance of the circuit will therefore be, per unit of length,

$$L_{s_1} = 2\left(2\mu \log_e \frac{d}{r} + \mu'\right). \quad . \quad . \quad . \quad . \quad . \quad (1)$$

In the case of copper conductors suspended in air we have practically $\mu = \mu' = 1$, whence

$$L'_{s_1} = 2\left(2 \log_e \frac{d}{r} + 1\right). \quad . \quad . \quad . \quad . \quad . \quad (2)$$

It will be noted that the logarithmic expression, giving the value of the flux in the space between the two conductors, shows that the induction decreases very rapidly as we get further from these latter; that is, beyond a certain distance between the conductors, the flux is not sensibly increased by augmenting that distance.

We can therefore say that in a circuit of any form whatever, the self-inductance is proportional to the length of the conductors composing the circuit, provided that these conductors be sufficiently far apart.

The part of the coefficient due to the sectional dimensions of each conductor is simply expressed by μ'. This quantity is negligible when, μ being equal to μ', the distance d is great compared to r, or else when the permeability μ is very great compared to μ'. We have an example of the former in the case of two copper wires stretched in the air, like telegraph-wires. The second case occurs in a copper conductor wound round an iron core of large diameter.

The above reasoning shows that the flux traversing the mass of conductors increases from their axis to their periphery. This way of reasoning takes for granted that the initial spreading of the current is uniform throughout the conductor's cross-section. If the current is variable, the effect of the unequal spreading of the flux over the cross-section is to create E. M. Fs. of induction which tend to increase the density of the current towards the periphery and, consequently, to still further increase the flux in this region. The consequence of these reactions is an increase of resistance; and if the frequency of the current is sufficiently great, the surface layers of the conductor alone are concerned in the electric displacement.

This effect is very clearly shown when a condenser is repeatedly discharged into a thick copper wire : an incandescent lamp can be made to glow, when put in shunt around a very small length of the wire, by reason of the difference of potential caused by the resistance of the external layers of the conductor, which are the only ones affected. Under these circumstances only the surface of the wire is heated by the current, the interior being heated by conduction. This can be shown by causing a tolerably large current of high frequency to pass through a wire for a very brief time : the wire, after growing hot for a moment on its surface, rapidly grows cool by diffusion of the heat throughout its mass.

M. Potier has given the following formula to express the ratio between the resistance R_a of a round wire for alternating currents of period T, and the resistance R_c of the same wire for continuous currents. Denoting by l the length of the wire, d its diameter, and ρ its specific resistance, we have

$$R_c = \rho \frac{4l}{\pi d^2},$$

and

$$\frac{R_a}{R_c} = 1 + \frac{1}{12}\frac{l^2}{R_c^2}\left(\frac{2\pi}{T}\right)^2 - \frac{1}{180}\frac{l^4}{R_c^4}\left(\frac{2\pi}{T}\right)^4 + \cdots$$

With a frequency of 80, the increase of resistance is 2.5 per cent. for a wire 15 mm. in diameter; it is 8 per cent. at a frequency of 133. The resistance R_a calculated by the above formula is that which, multiplied by the square of the effective current, gives the power lost in the Joule effect. It will be seen that, in order to avoid excessive cost of plant and working, conductors for strong alternating currents must have the form of tubes.

ROTATIONS UNDER THE ACTION OF INDUCED CURRENTS.

209. The reaction of the induced currents, in accordance with Lenz' law, enables us to obtain various movements of rotation.

Arago had noticed that if a copper disc rotates underneath a magnetized needle placed on a pivot over the centre of the disc, the needle is drawn in the direction of the disc's movement.

Likewise, if a strong magnet be rotated above a movable metallic disc, the latter tends to take up the same motion as the magnet. Referring to the explanation given in § 205, we see that the currents induced in front of the poles are repelled by the solenoidal currents equivalent to the magnet. The currents induced behind the poles are, on the contrary, attracted.

This experiment enables us to formulate a general rule:

Whenever the lines of force of a magnetic field are displaced in a conductor, it tends to follow the movement of the field.

210. Ferraris' Arrangement.—We are indebted to Ferraris for some ingenious arrangements based on the application of this rule.

Suppose a conductive disc movable about its axis O and

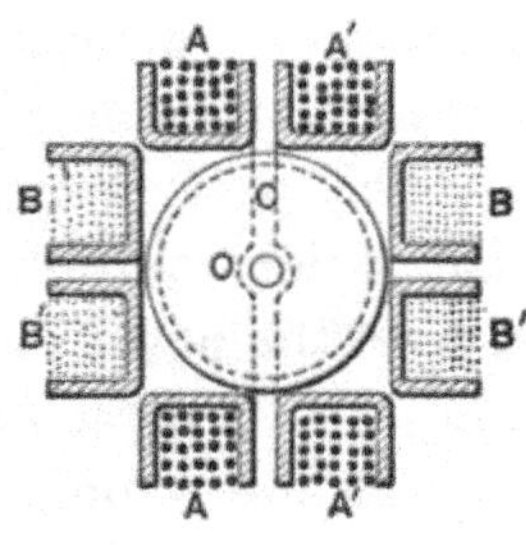

FIG. 94.

enclosed between two pairs of coils AA', BB', shown in section in the figure.

Similar periodic currents are passed through these coils, but differing in phase by $90°$.

A part of the lines of force thus produced traverses the substance of the disc, entering by the edge. Denote by $\mathcal{H}$ the mean intensity of the field formed, at any moment, in the disc, by the coils AA' and directed along the axis of these coils. $\mathcal{H}_m$ being the maximum intensity, we get

$$\mathcal{H} = \mathcal{H}_m \sin at.$$

If we suppose that the coils BB' produce a field whose

maximum intensity is also $\mathcal{H}_m$, but lagging a quarter-period behind, its mean intensity directed normal to $\mathcal{H}$ is

$$\mathcal{H}' = \mathcal{H}_m \sin\left(at - \frac{\pi}{2}\right) = -\mathcal{H}_m \cos at.$$

The two fields are compounded in the direction of a resultant which has at each instant an intensity

$$\mathcal{H}'' = \sqrt{{}^2\mathcal{H} + \mathcal{H}'^2} = \mathcal{H}_m.$$

We see that the resultant intensity is constant in magnitude, but that its direction varies. For $at = 0$, $\mathcal{H}''$ has the same direction as $\mathcal{H}'$; when at increases from 0 to $\frac{\pi}{2}$, the resultant field turns about O and becomes coincident with $\mathcal{H}$. For values of at comprised between $\frac{\pi}{2}$ and π the direction of the resultant performs another quarter-revolution; it returns to its initial position for $at = 2\pi$.

To sum up, the magnetic field produced by the combined action of the two pairs of coils revolves about O. The conductive disc is then traversed by induced currents whose reaction urges it to turn in the same direction as the field.

If the two pairs of coils were traversed by currents of the same phase, the direction of the resultant field would remain constant in the plane which bisects all the coils, but the intensity would vary from

$$+\mathcal{H}_m \sqrt{2} \quad \text{to} \quad -\mathcal{H}_m \sqrt{2}.$$

For every difference of phase between 0° and 90°, a rotating field will be obtained; the extremity of the straight line representing the resultant will describe, not a circumference as in the first case, but an ellipse whose major axis will lie along the bisecting plane.

211. Shallenberger's Arrangement.—Shallenberger has invented an arrangement which enables an analogous rotation to be obtained by the aid of a single periodic current, and this he has applied in the design of an alternating-current meter.

A coil of oblong form, traversed by a periodic current, surrounds a second smaller coil whose axis is inclined at about 45° to that of the larger coil. If we denote by L_m their mutual inductance, the E. M. F. induced in the second coil is

$$e' = - L_m \frac{di}{dt}.$$

This expression shows that the induced E. M. F. is $\frac{1}{4}$ phase behind the inducing current. In consequence of self-induction, the induced current likewise lags behind the E. M. F., so that the phases of the inducing and the induced currents differ by an angle comprised between 90° and 180°. Hence it follows that inside the coils is the seat of a rotatting magnetic field, which carries with it in its rotation a disc movable about a vertical axis. The effect is increased by using a disc of iron whose permeability increases the intensity of the field.

212. Repulsion Exercised by an Inducing Current on an Induced Current.—If we represent, by the graphic method of § 199, two neighboring currents lagging one behind the other by a quarter period, it is easy to see that during one half of their phase they are in the same direction and attract each other; during the other half, they are in opposite directions and repel each other: the resultant of the attractive forces is, moreover, equal to that of the repulsive forces, so that their mutual action may give rise to vibrations, but there will be no resultant effort of translation. We have just seen that the inducing current is always

more than 90° ahead of the secondary current on account
of the time-constant of the secondary circuit : such is the
case for the currents shown by full lines in Fig. 95.

The repellant period during which the currents are in
opposite directions has then a greater duration than the
period of attraction ; moreover the figure shows that during

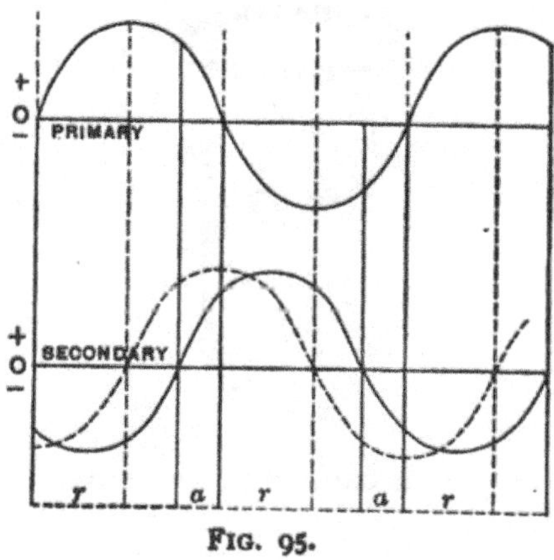

FIG. 95.

the period of repulsion the mean currents are greater than
during the attraction. It follows that a periodic inducing
current repels the induced current with a force which is
greater as the self-induction of the secondary circuit becomes greater.

Prof. Elihu Thomson has based some interesting experiments on these observations. If, for example, we put a
metallic ring *A* around the upper end of a straight electromagnet *B* traversed by alternating currents, it is found that
the ring is repelled to a position *A'*. On interposing
a metallic plate between the magnet and the ring, the ring
is observed to be no longer repelled. This is because the
screen itself becomes the seat of an induced current whose
effect on the ring neutralizes that of the primary current.
We shall see in the following chapter that such a screen

intercepts the electromagnetic waves through the action of which induction-effects are produced.

When we consider the repulsive effects with regard to the inducing periodic magnetic field, we observe that the circuit which is repelled tends to place itself so that the periodic flux traversing it shall be a minimum, for then the

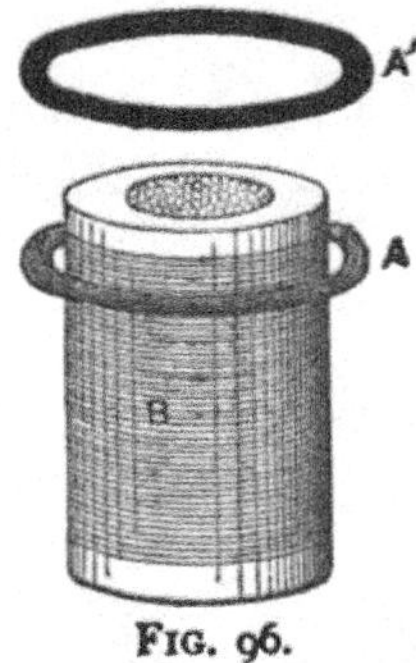

FIG. 96.

induced currents which cause the movement are reduced as much as possible.

The curious arrangements made by Ferraris and Prof. Elihu Thomson contain the germ of alternating-current motors.

Prof. Fleming has made an apparatus for measuring periodic currents, in which use is made of the phenomena of repulsion above described. A copper disc D (Fig. 97) is hung at the centre of a coil C, traversed by the undulatory currents.

The disc, being placed at aa' at an angle of 45° with the coil, is the seat of induced currents, and, in consequence of the repulsion exercised by the primary current, tends to place itself in the position bb', for which the periodic flux traversing it is a minimum. The mirror M enables us to read by reflection the angle at which the disc comes to rest, under the influence of the electrodynamic forces on one hand and the torsion of the suspension-wire on the other.

The disc must be placed obliquely to the coil, for if it were parallel to the turns of the coil it would oscillate con-

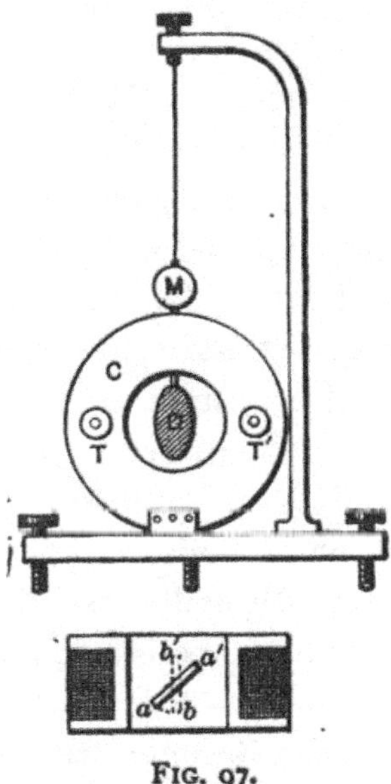

Fig. 97.

tinuously, the current giving it a new impulse at each crossing of the plane normal to bb'.

IMPEDANCE.*

213. Inductance.—A sinusoidal E. M. F. may be expressed by

$$e = E \sin \omega t,$$

where e is the instantaneous E. M. F. in volts;

      E is the maximum cyclic E. M. F. in volts;

      ω is the angular velocity of the E. M. F. in radians per second, and is $2\pi n$, where n is the number of cycles of the E. M. F. per second;

      t is the time expressed in seconds, starting from some suitable instant at which E is zero and changing from negative to positive.

When such an E. M. F. is impressed upon a circuit having a resistance of R ohms and an inductance of L henrys, a certain current strength will pass through the circuit. We assume that the inductance is non-ferric, that is to say, that the circuit is not linked with, or situated in the neighborhood of, iron, so that if there are coils of wire in the circuit, these coils are without cores. Then the current in the circuit will also be sinusoidal. This may be shown as follows:

The counter E. M. F. (C. E. M. F.) in the circuit at any instant due to the action of the inductance will be $-L\dfrac{di}{dt}$ volts, where i is the instantaneous current strength in amperes. The apparent C. E. M. F. due to fall of potential

---

* By A. E. Kennelly.

in the ohmic resistances will be $- iR$ volts; and if no electrostatic capacity exists in the circuit, these will be the only sources of C. E. M. F. present. But the impressed E. M. F., e, must be equal and opposite at any and every instant to the total C. E. M. F., or

$$e = E \sin \omega t = iR + L\frac{di}{dt} \text{ volts.}$$

The solution of this differential equation within a constant which is negligible as soon as the current has become steady, is

$$i = \frac{E \sin\left(\omega t - \tan^{-1}\frac{\omega L}{R}\right)}{\sqrt{R^2 + L^2\omega^2}} \text{ amperes. } \quad . \quad . \quad . \quad (1)$$

If the circuit contains resistance only, or $L = 0$, this becomes

$$i = \frac{E \sin \omega t}{R} = \frac{e}{R} \text{ amperes. } \quad . \quad . \quad . \quad . \quad . \quad (2)$$

That is, the current i would be in phase with the E. M. F. and would have at each instant the value $\frac{e}{R}$. The introduction of the inductance into the circuit has altered both the magnitude of the current and its phase. The magnitude becomes

$$\frac{e}{R\sqrt{1 + \frac{L^2\omega^2}{R^2}}},$$

which is always less than $\frac{e}{R}$, while the phase has become retarded by the angle $\tan^{-1}\frac{L\omega}{R}$, which is always less than 90°.

This effect of inductance may be considered from two distinct points of view, namely :

1st, and fundamentally, we may regard the circuit as having simply a resistance R, but as being the seat of two separate although interdependent E .M. Fs., one being the impressed sinusoidal E. M. F., e volts, and the other being the self-induced E. M. F., $L\dfrac{di}{dt}$ volts.

$$\text{Since} \qquad i = I \sin \omega t \text{ amperes,} \qquad \ldots \qquad (3)$$

$$\frac{di}{dt} = I\omega \cos \omega t = I\omega \sin \left(\omega t + \frac{\pi}{2}\right), \qquad \ldots \quad (4)$$

$$\therefore L\frac{di}{dt} = IL\omega \sin \left(\omega t + \frac{\pi}{2}\right) \text{ volts,} \qquad \ldots \quad (5)$$

This shows that the self-induced E. M. F. has the maximum value $IL\omega$ volts, that the effective value (square root of mean square) is $\dfrac{I}{\sqrt{2}}L\omega$ volts, and that the phase of this E. M. F. is 90° is $\left(\dfrac{\pi}{2}\right)$ ahead of the current i.

If, therefore, we represent the counter E. M. F. due to the fall of potential (iR volts) at any instant, say for convenience at its maximum value, by the straight line OI,

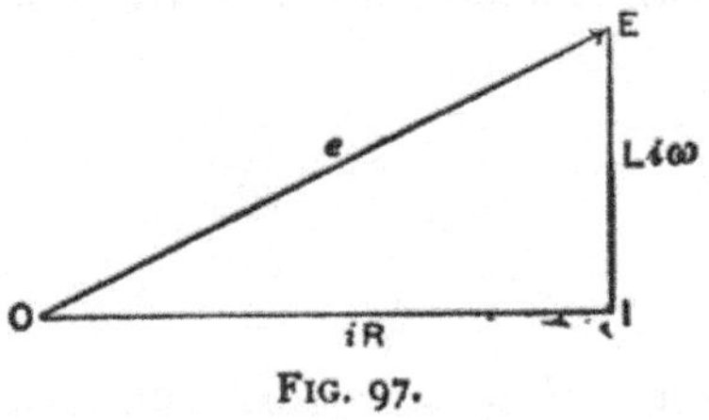

FIG. 97.

Fig. 97, then the product of this value into $\dfrac{L\omega}{R}$ will give the induced E. M. F. when turned through a right angle, or directed as shown by the line IE.

The geometrical sum of these two E. M. Fs. will be *OE* in both direction and magnitude, and this will be the impressed E. M. F., *e*, at the instant considered.

According to this conception, the E. M. F. *iR* which acts upon the resistance *R* so as to force through it the current *i*, is the geometrical difference *OI* of the impressed E. M. F. *OE* and the self-induced E. M. F. *IE*. In order to determine the current strength *i*, we have to compute the influence of *i* upon the C. E. M. F. of the circuit.

214. Inductance and Capacitance.—Extending this method to the case of a circuit in which a condenser of capacity *C* farads is connected with the sinusoidal E. M. F. *e* through an inductant resistance of *R* ohms and *L* henrys, we find the C. E. M. F. of self-induction to be $- L\dfrac{di}{dt}$ volts, and the C. E. M. F. of capacity to be $- \dfrac{1}{C}\displaystyle\int_{0}^{} idt$ volts, while the apparent C. E. M. F. of resistance is $- iR$ volts. The total C. E. M. F. in the circuit is consequently

$$- \left(iR + L\frac{di}{dt} + \frac{1}{C}\int_{0}^{t} idt \right) \text{ volts,}$$

so that, equating the impressed and reversed C. E. M. F., we have

$$e = E \sin \omega t = iR + L\frac{di}{dt} + \frac{1}{C}\int_{0}^{t} idt \text{ volts,} \quad . \ (6)$$

The solution of this equation is

$$i = \frac{E \sin\left\{ \omega t - \tan^{-1}\left(\dfrac{L\omega}{R} - \dfrac{1}{CR\omega} \right) \right\}}{\sqrt{ R^2 + \left(L\omega - \dfrac{1}{C\omega} \right)^2 }}. \quad . \quad . \ (7)$$

The effective current strength will, therefore, be

$$\dot{I} = \frac{E}{R\sqrt{2}} \frac{1}{\left\{ 1 + \dfrac{\left(L\omega - \dfrac{1}{C\omega}\right)^2}{R^2} \right\}} \text{ amperes.} \quad . \quad . \quad (8)$$

Plotting this case graphically in Fig. 98, OI is the

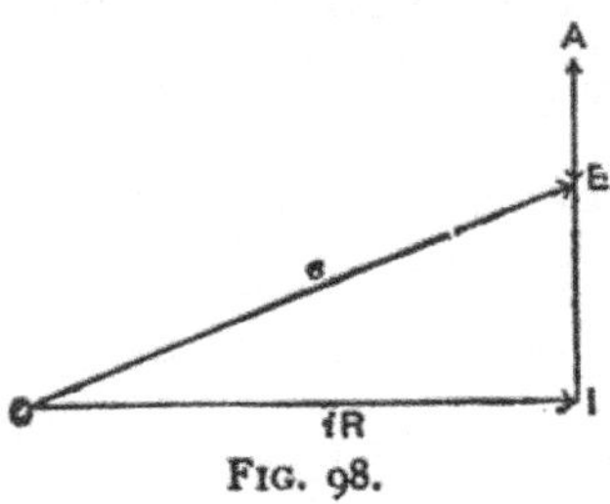

FIG. 98.

E. M. F. iR, active in producing the current i amperes, through the resistance R ohms. IA is the E. M. F. active in overcoming the C. E. M. F. of self-induction, and AE is the E. M. F. exerted in overcoming the C. E. M. F. of capacity, so that EA is the C. E. M. F. of capacity, while OE is the impressed E. M. F.

It is evident from the preceding that even with so simple a circuit as that comprising a sinusoidal E. M. F., resistance, inductance and capacity, the differential equation which has to be solved in order to evaluate the current strength is already formidable, and when any further degree of complexity is introduced, such as branch circuits, the difficulty of effecting the solution of the problem becomes very great.

215. Inductance and Reactance.—2d. The alternative method of dealing with sinusoidal alternating-current circuits is to assume that no E. M. Fs. act in the circuits other than the impressed E. M. Fs. In other words, the C. E. M. Fs. of inductance and capacity are ignored, but their effects are considered as resistances added to the circuit and opposed to the impressed E. M. Fs.

Thus, an inductance of L henrys is considered as a *reactance* of $L\omega$ ohms, and any capacity of C farads is considered as a reactance of $-\dfrac{1}{C\omega}$ ohms.

A reactance is reckoned in a direction making an angle of 90° with that of resistance, considered as a definite straight line or axis.

Thus, if a resistance of R ohms, Fig. 99, be connected in

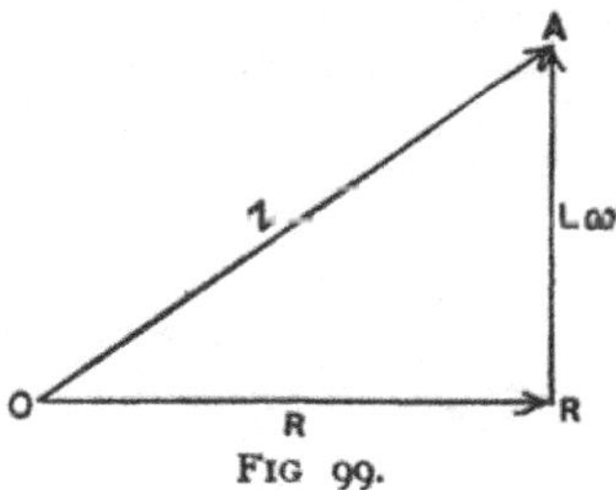

FIG 99.

circuit with an inductance of L henrys, the reactance of the latter will be $L\omega$ ohms, and the combined resistance and reactance will be $OR + RA$, where OR is the resistance and RA the reactance. Consequently, OA will be the impedance, or geometrical sum of OR and RA. If j denotes $\sqrt{-1}$, or the quadrantal versor operator, $RA = jL\omega$, and the impedance OA

$$Z = R + jL\omega \text{ ohms.} \qquad \qquad (9)$$

The effective current strength in the circuit will be in amperes

$$= \frac{\text{impressed effective E. M. F. in volts}}{\text{impedance in ohms}}.$$

$$\bar{I} = \frac{\bar{E}}{R + jL\omega} = \frac{\bar{E}}{\sqrt{R^2 + L^2\omega^2}}, \qquad (10)$$

where $\bar{E}$ is the effective E. M. F.

From the point of view of impedance, therefore, we consider a sinusoidal current circuit as acted upon by impressed E. M. Fs. only, and consider the resistance as modified by the presence of inductance or capacity.

Taking again a circuit of inductance, resistance and capacity,

The resistance is R ohms;

The reactance of inductance is $jL\omega$ ohms;

The reactance of capacity is $-\dfrac{j}{C\omega}$ ohms;

The impedance is $R + j\left(L\omega - \dfrac{1}{C\omega}\right)$ ohms;

so that the current strength is

$$\bar{i} = \frac{\bar{E}}{R + j\left(L\omega - \dfrac{1}{C\omega}\right)} \text{ amperes.} \quad . \quad . \quad . \quad (11)$$

This expression is equivalent to that of (8).

When a sinusoidal current circuit contains any number of inductances, resistances and condensers in series, the impedance is the geometric sum of all the resistances and reactances. That is,

$$\dot{I} = \frac{\bar{E}}{\Sigma R + j\left(\Sigma L\omega - \Sigma\dfrac{1}{C\omega}\right)} \text{ amperes.} \quad . \quad . \quad (12)$$

The computation may be carried out by algebraic rules or by geometry.

It is to be observed that the above equation is a versor equation, and that i is no longer a simple number of amperes, but a number of amperes turned through a certain angle. The execution of the operations demanded by equation (12), namely, the division of the quantity E by the complex quantity representing the impedance, results in a quotient $\bar{I}$, which is also a complex quantity and should be represented by a line having both direction and magnitude, the direction being confined to a single plane and to an angle less than 90° from E.

216. Joint Impedances.— When sinusoidal current circuits are connected in parallel, their joint impedance presents, symbolically, no more difficulty than the treatment of joint resistances in continuous-current circuits. Thus if Z_1, Z_2, $Z_3 \ldots Z_n$ be the impedances of n circuits connected in parallel, each impedance being a plane vector, or versor quantity, expressible in the form

$$Z_n = R_n + j\left(L_n\omega - \frac{1}{C_n\omega}\right) \text{ ohms}, \quad . \quad . \quad . \quad (13)$$

then the reciprocal of Z_n may be called the *admittance* of the circuit n. The admittance of a circuit is a plane vector or directed quantity expressible in mhos. Thus if Y_n be the admittances of circuit n,

$$Y_n = \frac{1}{Z_n} = \frac{1}{R_n + j\left(L_n\omega - \frac{1}{C_n\omega}\right)} \text{ mhos.} \quad . \quad . \quad (14)$$

The joint admittance of the group of multiple circuits will then be the geometrical sum of all the admittances, or

$$Y = Y_1 + Y_2 + Y_3 + \ldots + Y_n \text{ mhos}, \quad . \quad . \quad (15)$$

$$= \frac{1}{R_1 + j\left(L_1\omega - \frac{1}{C_1\omega}\right)} + \ldots + \frac{1}{R_n + j\left(L_n\omega - \frac{1}{C_n\omega}\right)}. (16)$$

The joint impedance will be the reciprocal of the joint admittance, or

$$Z = \frac{1}{Y} = \frac{1}{Y_1 + Y_2 + \ldots + Y_n} \text{ ohms}.$$

Z will obviously be a plane vector quantity.

When any inductant resistance in a sinusoidal circuit is linked magnetically with a secondary circuit, through a mutual inductance of L_μ henrys, the presence of the secondary circuit will modify the impedance of the inductant resistance in a definite manner.

Let R_s and X_s be the resistance and total reactance of the secondary circuit ;

R_p and X_p be the corresponding resistance and reactance of the inductant resistance forming the primary; then if $n = \dfrac{L_\mu \omega}{Z_s}$, where

$$Z_s = R_s + jX_s,$$

the effective impedance of the inductant resistance acting as the primary is changed from $R_p + jX_p$ in the absence of the secondary circuit to

$$Z = R_p + n^2 R_s + j(X_p - n^2 X_s) \text{ ohms}$$

in the presence of the secondary circuit.

THE PROPAGATION OF CURRENTS.

GENERAL CONSIDERATIONS.

217. Phenomena which Accompany the Propagation of the Current in a Conductor.—Ohm has furnished an expression for a continuous current in a conductor, relying on an assimilation between the electric and the calorific fluxes.

If we call r the resistance of a conductor per unit length, $-\dfrac{dU}{dl}$ the variation of potential per unit length in the direction l of the conductor, we have

$$-\frac{dU}{dl} = ri. \quad\quad\ldots\ldots\ldots \quad (1)$$

Lord Kelvin has carried this law further so as to make it applicable to the variable period of a current which is being set up in a conductor of given capacity. Imagine an insulated cable submerged in water. This cable forms a cylindrical condenser, with the water, supposed to be at zero potential, as its outer plate. The moment a difference of potential is established between the two ends of the conductor, a current is set up; but at the same time every portion of the conductor is charged with a quantity of electricity in proportion to its capacity and the difference between the potentials of the plates. There then occurs, across the dielectric of the cable, a charge, or *displacement-current* (Maxwell), which may be considered as a derived current from that in the conductor. This displacement-current ceases when the tension of the dielectric balances the potential-difference of the condenser-plates.

Then there exists in the dielectric an electric field whose lines of force connect the condenser-plates and end in equal and opposite quantities of electricity (§ 88).

When the period of charge is past, there only remains the permanent current expressed by equation (1).

The same phenomenon occurs, but in a lesser degree, when a current is sent through a circuit in the air, for the charge in the conductor excites a contrary charge upon the neighboring conductors separated by air or other dielectric.

Denote by c the capacity of the conductor per unit of length; the corresponding charge for a difference of potential U is

$$q = cU. \quad . \quad . \quad . \quad . \quad . \quad . \quad . \quad (2)$$

The energy of the electric field per unit length is $\frac{1}{2}cU^2$.

Let us express the fact that the quantity of electricity which enters the given segment of the cable is equal to that which flows out, plus the current across the dielectric. The variation of the current in the conductor per cm is represented by $-\dfrac{di}{dl}$; the displacement-current is $\dfrac{dq}{dt}$, whence the condition

$$-\frac{di}{dl} = \frac{dq}{dt} = c\frac{dU}{dt}. \quad . \quad . \quad . \quad . \quad (3)$$

Combining (1) and (3), we get the equation

$$\frac{d^2U}{dl^2} = rc\frac{dU}{dt}. \quad . \quad . \quad . \quad . \quad . \quad (4)$$

This elementary law has enabled Lord Kelvin to investigate the variable period in submarine cables, where condenser-phenomena play an all-important part.

It must be noted that the current traversing the conductor is manifested by a loss of energy transformed into heat by the Joule effect.

The displacement-current in the dielectric represents, on the contrary, a storage of energy in the potential state, shown by the tension of the dielectric. This energy in its turn produces heating effects when the cable is discharged.

In order to investigate the manifestations, observed during the variable period, in a general way, it is necessary to call in the magnetic phenomena produced by the current in the surrounding medium.

We have just seen that the current develops an electric field whose intensity depends on the specific inductive capacity of the dielectric and whose lines of force are normal to the conductor. But besides this, the current creates a magnetic field characterized by lines of force forming closed curves around the conductor; the intensity of this field is proportional to the permeability of the surrounding medium magnetized by the current.

The intensity of the magnetic field diminishes rapidly with the distance from the conductor. We may therefore say that the self-induction of a circuit is sensibly proportional to the length of the conductors composing it, on condition that they are far enough apart so that the lines of force developed by them do not encroach on each other.

Such an encroachment occurs in the case of two wires stranded together and connected in series. They tend, under the influence of an electric current, to produce lines of force opposite in direction in the surrounding medium, so that their coefficient of self-induction is practically zero. The same holds for a bobbin wound with a wire bent double (§ 188).

Putting these cases aside, denote by L, the coefficient of self-induction of a circuit per unit length of the conductors.

The magnetic energy of the current, represented by the magnetization of the medium, is $\frac{1}{2}L_i i^2$ per centimetre.

The medium opposes a certain inertia to magnetization, which has the effect of developing an E. M. F. contrary to that which gives rise to the current. This E. M. F. is $-L_i\frac{di}{dt}$ per centimetre, so that Ohm's formula when completed becomes

$$-\frac{dU}{dl} = ri + L_i\frac{di}{dt}. \quad . \quad . \quad . \quad . \quad . \quad . \quad (5)$$

Equations (3) and (5) enable us to treat the problem of the variable period in its whole extent, taking into account the production of the electric and the magnetic fields created by the current.

We have supposed the current to be surrounded by a perfect insulator. If there were any losses of electricity across the dielectric, we would have to add to the second member of the preceding equation a term accounting for the derived currents due to the electric conductivity of the medium.

Equations (3) and (5) are analogous to those met with in the theory of the propagation of sound-waves, when we admit that the passive resistance of the medium is proportional to the first power of the velocity, and, moreover, that electric resistance corresponds to friction, self-induction to the inertia of the medium, and capacity to the reciprocal of a pressure.

From this analogy it follows that, if a circuit be subjected to a periodic E.M.F., the electric waves generated are propagated in accordance with laws identical with those of the propagation of sound. In particular, if a periodic E. M. F. is applied to one of the ends of a line insulated at the other end, the electric waves thus created are reflected at the insulated end and return to their starting-point, where they are reflected again, just like sound-waves sent into a tube which is closed at one end.

The analogy noticed above is of particular interest in telephony, for it shows that the electric waves transmit speech according to laws identical with those which govern its propagation in a ponderable medium.*

218. Special Characteristics shown by Alternating Currents.—As has been shown in § 199, alternating currents do not add together as do continuous currents, but are compounded like vectors according to the parallelogram of forces: it is for this reason that two equal periodic E. M. Fs. acting in a circuit do not usually give a resultant current double that which could be produced by each one of them. The mean resultant is not equal to the sum of the mean component currents unless the phases of the latter are together; it is zero if the phases differ by 180°, just as the resultant of two forces is zero when they are equal and in opposite directions.

This view explains certain peculiar effects produced by periodic electromotive forces.

Consider, for example, an alternating current split into two branches having different resistances and self-inductions.† The branch currents will present differences of phase with regard to each other and to the total current, § 198; at each instant the total current will be equal to the sum of the branch-currents, but the *mean* total current will by no means be equal to the sum of the *mean* branch-currents. If the difference of phase of the latter is great enough, it may even happen that the mean current of each one of the branches will be greater than the whole current.

---

* See Demany, *Théorie de la propagation de l'électricité.* Bulletin de L'association des ingénieurs sortis de l'Institut Montefiore, 1890.

† Lord Rayleigh, *On Forced Harmonic Oscillations.* Phil. Mag., May, 1886.

It is sufficient, for this purpose, to have the angle of lag more than 120°, for the parallelogram of forces shows that when two equal vectors make an angle of 120°, their resultant is equal to one of the component vectors. The sine-curve representing the total current will have, at each instant, its ordinates equal to the algebraic sum of the ordinates of the two component sine-curves. These latter may therefore have very much greater coördinates of the resultant curves.

Take another case, pointed out by M. Smith. Between two points a and d subjected to an alternating potential-difference of constant mean value, put two resistances ab and bd in series, one having a considerable coefficient of self-induction, the other non-inductive. On each resistance put a glow-lamp on a shunt. The two lamps are alike, and the non-inductive resistance is adjusted until they both burn with the same brilliancy; this result shows that equal currents are traversing the filaments, and that the potential-difference of the points a, b is the same as that of b, d. Next, the wire joining the lamps is separated from that connecting the resistances, and the brilliancy of the lamps is seen to decrease, although the potential-difference of the points a, d retains the same mean value; which proves that this latter is less than the sum of the mean differences found between the points a, b and b, d. This fact is also explained by a difference of phase in the potential-differences to which the two resistances in series are subjected: as the resultant has remained constant, the two components, whose phases are not together, must have assumed values greater than half the acting potential-difference.

Instead of using glow-lamps to indicate potential-differences, we might have used the quadrant electrometer as arranged in § 102 (II).

It is unnecessary to state that there is no contradiction of the principle of the conservation of energy in these experiments. If we find an increase of power consumed in one branch of a circuit, we observe a corresponding expenditure at the source of electricity, by the increase in the mean power of this source,

$$P_m = E_{eff} \cdot I_{eff} \cdot \cos \phi.$$

219. Comparative Effects of the Self-induction and Capacity of a Circuit.—It is interesting to observe that in considering the flux of electricity which flows during the variable period, the capacity and self-induction play opposite parts. In fact, the displacement-current due to the capacity is added to the current which traverses the conductor, so that the phenomenon of condensation is equivalent to an apparent diminution of the resistance of the circuit during the variable period.

Suppose a conductor of resistance r, whose ends are at a potential, one of U, the other zero. Insert, between the extremities of the conductor, a condenser of capacity c; its charge at the end of the variable period is $q = cU = cir$. This charge is added to the quantity of electricity traversing the circuit.

On the contrary, the electromagnetic induction produces an increase of the impedance and a diminution of the flux of electricity, during the variable period of closure, equal to (§ 194)

$$\frac{\Phi}{r} = \frac{L_s i}{r} = q'.$$

From these opposite effects follows a certain compensation which can be made use of in the transmission of signals in cables.

The difference of the fluxes of the extra-current and displacement-current is

$$q' - q = \frac{L_s i}{r} - cir = \frac{i}{r}\left(L_s - cr\right).$$

We see that, as regards the flux of electricity transmitted during the variable period, the effect of the condensation corresponds to a diminution of the self-induction equal to the product of the capacity by the square of the resistance of the conductor.   This observation is due to Sumpner.

220. Effect of a Capacity in a Circuit Traversed by Alternating Currents.*—A condenser may be inserted in series in a circuit traversed by alternating currents without interrupting the passage of these currents, as would be the case if we were dealing with a continuous E. M. F.   In fact, at each reversal of the current the condenser is discharged, and recharged in the opposite direction.   It is necessary, however, if a powerful mean current is desired, that the capacity of the condenser should be great enough to absorb the electric flux transported by the waves of the current.

Suppose that in a circuit of resistance r and without self-induction, we insert a condenser of capacity c.

Denoting by $E = E_0 \sin at$ the periodic E. M. F., and by u the potential-difference of the plates at an instant t, we have

$$E = E_0 \sin at = u + ri. \quad . \quad . \quad . \quad . \quad . \quad (1)$$

But

$$cdu = idt. \quad . \quad . \quad . \quad . \quad . \quad . \quad (2)$$

Differentiating (1) and substituting for du its value obtained from (2), we arrive at

$$aE_0 \cos at\, dt = \frac{i}{c} dt + rdi. \quad . \quad . \quad . \quad (3)$$

This equation has a form similar to equation (3) of § 180. Resolving it by the same method and suppressing the exponential term, which becomes zero at the end of a very

---

short time, we find for the expression of the regular current

$$i = \frac{E_{\text{o}}}{\sqrt{r^2 + \dfrac{1}{a^2 c^2}}} \sin (at + \phi) \quad \ldots \quad (4)$$

on condition that

$$\phi = \tan^{-1} \frac{1}{acr}. \quad \ldots \ldots \quad (5)$$

This result shows that the capacity has the effect of *advancing* the phase of the current ahead of the E. M. F.

We have moreover

$$I_{\text{eff.}} = \frac{E_{\text{eff.}}}{\sqrt{r^2 + \dfrac{1}{a^2 c^2}}} < \frac{E_{\text{eff.}}}{r};$$

which proves that the condenser reduces the current the more, the less its capacity is. The current becomes zero for $c = 0$. An infinite capacity would produce the same effect as taking the condenser away.

We could have found the preceding equations directly by substituting, in the equations of § 180, $-\dfrac{1}{ac}$ for the factor aL_t. This fact is explained if it is observed that the self-induction introduces an electromotive force $e = -L_t \dfrac{di}{dt}$, while a condenser brings in a difference of potential (taken from (2))

$$e' = -u = -\frac{1}{c} \int^t i \, dt.$$

Substituting for i its value $I \sin at$ in these expressions, we find

$$e = -aL_t I \cos at,$$

$$e' = +\frac{1}{ac} I \cos at;$$

values which are equal if we put $aL_t = -\dfrac{1}{ac}$.

221. Combined Effects of a Capacity and a Self-induction in a Circuit Traversed by Alternating Currents. Ferranti Effect. — If we introduce in a circuit traversed by alternating currents a self-induction and a capacity, as the first tends to make the phase of the current lag, and the second tends to advance it, there will be a more or less complete neutralization of the two effects.

Let us analyze this combination: Keeping the preceding notation, we will have the two equations

$$E = E_0 \sin at = L_s \frac{di}{dt} + ri + u, \quad \ldots \quad \text{(1)}$$

$$cdu = idt, \quad \ldots \quad \ldots \quad \text{(2)}$$

and, combining them,

$$L_s c \frac{d^2i}{dt^2} + rc \frac{di}{dt} + i - E_0 ac \cos at = 0. \quad \ldots \quad \text{(3)}$$

The general solution of this differential equation is of the form

$$i = Ae^{mt} + B \sin at + D \cos at. \quad \ldots \quad \text{(4)}$$

Differentiating this equation twice and introducing in (3) the values of $i \frac{di}{dt}$ and $\frac{d^2i}{dt^2}$ thus found, we can determine the arbitrary coefficients m, B, and D. If we notice moreover that the exponential term is very quickly zero with t, the value of the established current i reduces to

$$i = \frac{E_0}{\sqrt{r^2 + \left(aL_s - \frac{1}{ac}\right)^2}} \sin (at - \phi) \quad \ldots \quad \text{(5)}$$

on condition that

$$\phi = \tan^{-1} \frac{aL_s - \frac{1}{ac}}{r}. \quad \ldots \quad \ldots \quad \text{(6)}$$

The effective current is

$$I_{\text{eff.}} = \frac{E_{\text{eff.}}}{\sqrt{r^2 + \left(aL_s - \dfrac{1}{ac}\right)^2}}. \quad \cdots \cdots \quad (7)$$

This expression could have been found directly by supposing a circuit with two self-inductions L_s and L_s' and substituting for the second one a capacity such that

$$aL_s' = -\frac{1}{ac}.$$

Equation (6) shows that the current-phase will be behind or ahead of the E. M. F. phase according as aL_s is larger or smaller than $\dfrac{1}{ac}$, or $a^2 L_s c \lessgtr 1$. In every case the current will be less than if there were neither self-induction nor capacity, unless $a^2 Lc = 1$.

If we determine from equation (5) the expression for the E. M. Fs.

$$e = -L_s \frac{di}{dt} \quad \text{and} \quad e' = -\frac{1}{c}\int i\,dt$$

due to the self-induction and the capacity, we find

$$e = -L_s \frac{aE_0}{\sqrt{r^2 + \left(aL_s - \dfrac{1}{ac}\right)^2}} \cos(at - \phi),$$

$$e' = +\frac{1}{ac} \frac{E_0}{\sqrt{r^2 + \left(aL_s - \dfrac{1}{ac}\right)^2}} \cos(at - \phi).$$

It will be noted that these values may be very much higher than E; in particular, if $aL_s = \dfrac{1}{ac}$ and r tends towards 0, it is easily seen that the potential differences at the terminals of the induction-coils and the condenser tend towards infinity. These peculiarities are explained by the composition of the E. M. Fs., as has been seen in § 218.

It is to a similar cause that the effect must be attributed which was observed in Ferranti's cables in London : when these cables, which are concentric and have a considerable capacity, are traversed by alternating currents, an elevation of tension is observed at the end of the line.

From all the above deductions it follows that the capacity of a circuit corrects the effects of the self-induction by decreasing the impedance created by the latter, as well as the phase-lag. The capacity to give a circuit in order to completely neutralize the effects of its self-induction is shown by the equation $a^2 L_s c = 1$. If $a = 100$, for example, and $L_s =$ one quadrant, we will have $c = 100$ microfarads. This capacity, which can only be obtained by the use of expensive condensers, could be reduced by artificially increasing the self-induction by electromagnets with closed magnetic circuits.

The problem of the subdivision of a periodic current in a branch having self-induction, and another with a condenser, also gives rise to interesting observations. The current in the inductive resistance will have a phase-lag behind the resultant current, while the flux in the condenser will be advanced. Consequently the sum of the two derived fluxes will be greater than the resultant flux, and each of the branches may even be traversed by a mean current superior to the total current.

222. Oscillating Discharge.—Let us take up again the discharge of a condenser, by investigating which we approached the study of the electric current (§ 103), being guided by our knowledge of the laws of induction.

Suppose a condenser of capacity c, whose plates are at a difference of potential U. Call r and L_s the resistance and self-induction of the circuit of discharge. It is well to note that r expresses the metallic resistance of the discharge-circuit.

The discharge-current is equal to the rate of variation of the charge, or

$$i = -\frac{dq}{dt} = \frac{U - L_s\frac{di}{dt}}{r};$$

but $q = cU$, whence

$$i = \frac{\frac{q}{c} + L_s\frac{d^2q}{dt^2}}{r},$$

and

$$\frac{d^2q}{dt^2} + \frac{r}{L_s}\frac{dq}{dt} + \frac{q}{cL_s} = 0.$$

To resolve this equation, put $q = e^{mt}$, and we get

$$e^{mt}\left(m^2 + \frac{r}{L_s}m + \frac{1}{cL_s}\right) = 0.$$

The general integral is of the form

$$q = Ae^{m_1 t} + Be^{m_2 t}, \quad \cdots \cdots \quad (1)$$

A and B being constants of integration; m_1 and m_2, the roots of the equation obtained by equating to zero the trinomial in parenthesis, or

$$-\frac{r}{2L_s} \pm \sqrt{\frac{r^2}{4L_s^2} - \frac{1}{cL_s}}.$$

Substituting these values in (1) and putting $\frac{L_s}{r} = \tau$, we get

$$q = e^{-\frac{t}{2\tau}}\left(Ae^{\sqrt{\frac{1}{4\tau^2} - \frac{1}{c r \tau}}\cdot t} + Be^{-\sqrt{\frac{1}{4\tau^2} - \frac{1}{c r \tau}}\cdot t}\right). \quad . \quad (2)$$

If the roots of the trinomial are imaginary, the radical exponent takes the form

$$\sqrt{\frac{1}{c r \tau} - \frac{1}{4\tau^2}}\sqrt{-1}\cdot t,$$

and equation (2) reduces, by Euler's formula, to

$$q = e^{-\frac{t}{2\tau}}\left(M \cos\sqrt{\frac{1}{c r \tau} - \frac{1}{4\tau^2}} \cdot t + N \sin\sqrt{\frac{1}{c r \tau} - \frac{1}{4\tau^2}} \cdot t\right). \quad (3)$$

The constants of integration are determined by the simultaneous conditions

$$t = 0, \quad i = 0, \quad q = Q$$

introduced in equations (2) or (3) and in the expression

$$i = -\frac{dq}{dt}.$$

Then substituting in this last the values found, we get, in the case of real roots,

$$i = \frac{Q}{2 c r \tau \sqrt{\frac{1}{4\tau^2} - \frac{1}{c r \tau}}} e^{-\frac{t}{2\tau}}\left(e^{\sqrt{\frac{1}{4\tau^2} - \frac{1}{c r \tau}} \cdot t} - e^{-\sqrt{\frac{1}{4\tau^2} - \frac{1}{c r \tau}} \cdot t}\right). \quad (4)$$

In the case of imaginary roots, the value of the current assumes the form

$$i = \frac{Q}{c r \tau \sqrt{\frac{1}{c r \tau} - \frac{1}{4\tau^2}}} e^{-\frac{t}{2\tau}} \sin\sqrt{\frac{1}{c r \tau} - \frac{1}{4\tau^2}} \cdot t. \quad (5)$$

Equation (4) shows that, for

$$4\tau > c r \quad \text{or} \quad r < \sqrt{\frac{4L}{c}},$$

the discharge occurs in the form of a continuous current, constant in direction, beginning with a value of zero, rising rapidly to a maximum, and then rapidly decreasing.

When

$$r < \sqrt{\frac{4L}{R}},$$

the discharge-current oscillates periodically between positive and negative values which decrease rapidly.

Equation (5) shows that the oscillating current passes through the same phases for

$$\sqrt{\frac{1}{cr\tau} - \frac{1}{4\tau^2}} \cdot t = 0, \quad 2\pi, \quad 4\pi, \quad \ldots$$

Hence we deduce that the period of the oscillating current is

$$T = \frac{2\pi}{\sqrt{\dfrac{1}{cr\tau} - \dfrac{1}{4\tau^2}}} = \frac{2\pi}{\sqrt{\dfrac{1}{cL_s} - \dfrac{r^2}{4L_s^2}}}.$$

If r is negligible compared with $2\sqrt{\dfrac{L_s}{c}}$, we get

$$T = 2\pi \sqrt{cL_r}.$$

If, for example, $k = 1$ microfarad, $L_s = 1.10^{-4}$ quadrant, $r = 0$,

$$T = 2\pi \sqrt{1.10^{-6} \times 1.10^{-4}} = 0.000063 \text{ second.}$$

Figures 100, 101, and 102 give a graphical representation of the above phenomena.

The curve (I) represents the variations of the charge in terms of the time in the case of a continuous discharge, and

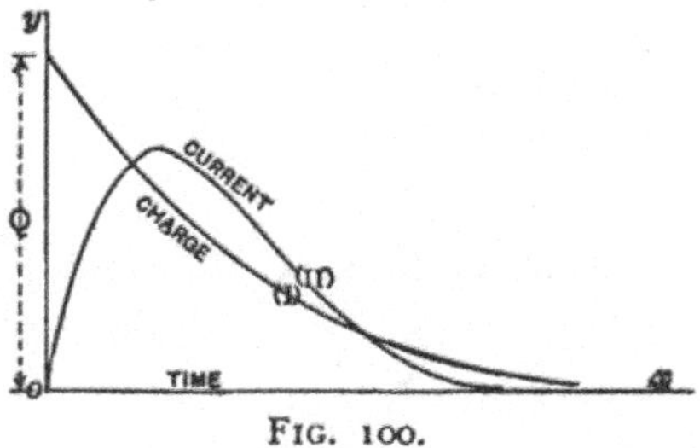

Fig. 100.

curve (II) shows the variations of the current which, zero at the beginning of the discharge, rapidly assumes a maximum value and then decreases asymptotically towards zero.

Fig. 101 shows the variations of the charge in the case of an *oscillating discharge*, and Fig. 102 shows the current-

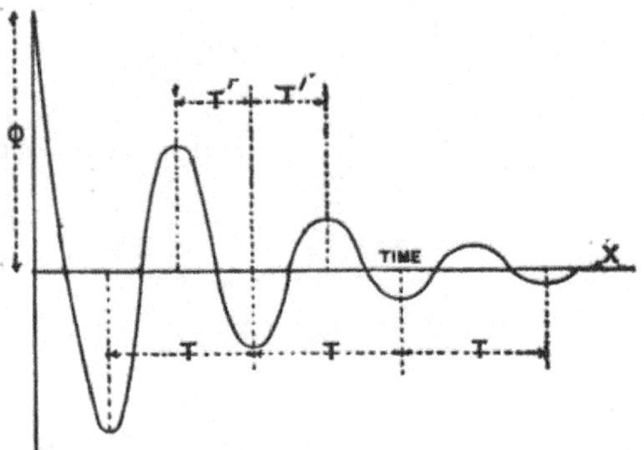

FIG. 101.

variations. These variations are shown by decreasing undulations whose period is equal to T.

We are indebted to Lord Kelvin for the first analytical

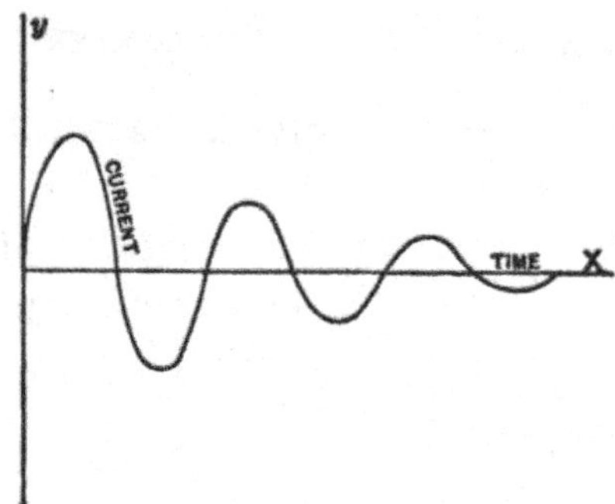

FIG. 102.

investigation of the oscillatory discharge, which of late years has been the subject of experiments performed by, among others, Hertz and Lodge.

To understand these phenomena, we must recall the fact that the dielectric of a charged condenser is subjected to a tension which may be compared to that of a spring. If the cause which produces the tension disappears suddenly, the dielectric returns to its original position after having performed oscillations comparable to those of a spring when suddenly released.

To keep the spring from oscillating we must oppose a resistance to its movement, by plunging it, for example, in a viscous fluid. Thus, too, by presenting an electrical resistance to the discharge of a condenser, it is rendered continuous.

The period of the oscillations of a spring depends upon its mass or its inertia. So the period of the electric discharge varies with the self-induction, which represents the magnetic inertia of the medium surrounding the circuit.

By increasing the self-induction we increase the length of the period more and more; to reach this result, it is sufficient to pass the discharge through a bobbin, the number of whose turns increases progressively. Curiously enough, it is useless to put an iron core in the bobbin, for, in consequence of the rapidity of the reversals of the discharge, the result is zero as far as affects the magnetization.

By proceeding in this way, *i.e.*, by interposing an increasing number of turns of wire between the condenser-plates, we get an oscillating discharge of increasing period. The discharge-spark, which occurs at every break in the conductors, appears single on account of the rapidity of the phenomenon, but, if it is reflected from a rotating mirror, we see that it seems composed of a succession of luminous points. On increasing the period sufficiently, the chain of luminous points becomes visible to an observer who looks at the spark through a glass which he moves rapidly. Finally, when the self-induction of the discharge-circuit and the capacity of the condenser are sufficiently great, the impulses communicated to the air by the electric undulations reach the limit of vibrations perceptible by the ear, and the spirals give a note whose height can be diminished at will.

Dr. Lodge, by suitably varying the values of r, k, $L_{,}$, has succeeded in obtaining a scale of electrical vibrations whose periods extended from one hundred-millionth to one five-hundredth of a second.

223. Transmission of Electric Waves in the Surrounding Medium.

—The comparison of the oscillations of an electric discharge with those of a vibrating elastic body can be carried still further. A tuning-fork generates sound-waves in the surrounding air, which spread out in space and can be shown to be present by means of a *resonator* tuning-fork in pitch with the first. If sound-waves hit perpendicularly against a solid wall, they are reflected; the incident waves interfere with those sent back from the wall, and nodes and loops of vibration are formed which follow each other alternately. The resonator remains mute at the nodes and marks the position of the loops loudly. The length of the wave is equal to double the distance between two consecutive nodes. The velocity of transmission of the waves is the quotient of the length of a wave by its duration.

Analogous phenomena appear in the propagation of calorific and luminous radiations, formed by vibrations transverse to the direction of the rays.

Dr. Hertz has experimentally demonstrated that discharge-oscillations between electrified bodies produce, in the surrounding medium, electric waves whose properties are identical with those of the radiations emitted by bodies at high temperatures, that is to say that they give rise to phenomena of reflection, interference, refraction, polarization, and diffraction. In order to exhibit these properties, it was necessary, first of all, to have an apparatus producing continual electric oscillations.

Dr. Hertz made use of a Ruhmkorff coil for this purpose, Fig. 103, the secondary terminals of which are connected to two conductors which constitute an electric *vibrator* or *exciter* and which have been given a number of shapes. In Fig. 103, the vibrator V is formed by two conductive rods, situated in the line of each other's prolongation and terminated at their adjoining ends by metallic knobs, the opposite ends carrying metal spheres. These latter can be replaced by discs or by metal plates hung from the rods. The induction-coil is intended to keep up the electric charges on the two parts of the vibrator. In virtue

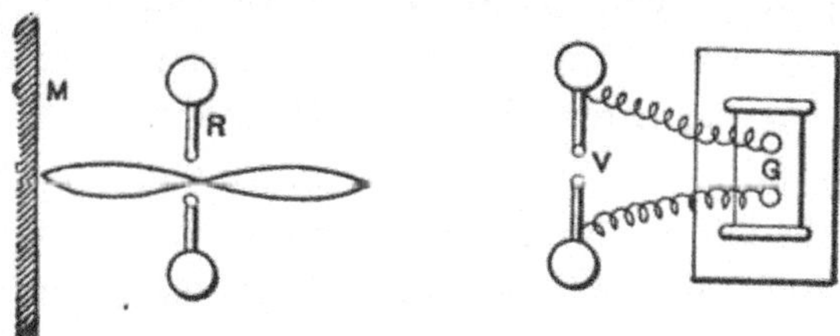

FIG. 103.

of these alternately opposed charges, the medium surrounding the vibrator is the seat of an electric field whose lines of force, periodically reversed in direction, connect the two parts of the apparatus. The mean direction of the field, that is to say of the electric forces, is the axis of the conductive rods. In consequence of the high potential of the charges, they recombine under the form of sparks leaping between the two metallic knobs. The oscillations of this prolonged discharge are dependent on the capacity of the vibrator and the resistance and self-induction of the conductors composing it: note that here the resistance in question is that of the conductors connected to the spheres. When the charge of these latter is sufficient, a spark occurs between the adjoining knobs, and it is only then that electric oscillations take their rise between each sphere and the corresponding knob. The duration of these oscillations is very small on account of their rapid decrease,

Fig. 101; the induction-coil must, therefore, renew the charges of the spheres very rapidly, and the sparks succeed each other at very short intervals.

The oscillations, which produce rapid alternating currents in the rods of the vibrator, are propagated in the surrounding medium.

To demonstrate this, Dr. Hertz made use of an apparatus capable of vibrating in unison with the *electric vibrator* and which is called an electric *resonator.*

The resonator can be identical with the vibrator, as in Fig. 103, or else it may have an entirely different form, provided that the three characteristic magnitudes, capacity, resistance, and self-induction of the conductors composing it, satisfy the same conditional equation as the corresponding quantities of the vibrator,—a fact which can also be experimentally verified.

This being so, if we place the resonator in the vicinage of the vibrator so that their axes are parallel, a flow of sparks is observed between the knobs of the first-named apparatus. These sparks, caused by the variations of the electric force parallel to the axis of the conductors, diminishes steadily as the resonator is moved further off.   The explosive distance naturally decreases with the maxima of the electric force.

The phenomenon of electric resonance is produced even if a solid insulating wall is interposed between the vibrator and the resonator, *e.g.*, a vertical partition of wood or masonry; but it ceases when the partition is a conductor. In this latter case, if the resonator be moved between the vibrator and the partition, it is found that in certain points the spark ceases and that in others it is re-enforced.

These extinctions, which are repeated at intervals of equal length, prove that the transmission of electric forces takes place in the form of *waves* capable of being reflected by a conductive wall, the waves sent back by the wall being capable of interfering with the incident waves to form the observed nodes and loops of vibration. The distance between two nodes corresponds to half a wave-length.

To obtain a number of nodes in the limits of the room used for experimenting in, it is necessary to produce short waves. By using as vibrator a brass tube 26 cm. long and 3 cm. in diameter, divided into two parts with the adjoining ends terminating in segments of spheres, Hertz succeeded in exciting waves only 30 cm. in length. The resonator was a straight wire 1 metre long, also divided in two parts and furnished with metallic knobs between which the sparking takes place.

Hertz has likewise performed a very interesting series of experiments with the aid of a metallic parabolic reflector designed to concentrate the electric undulations in a given direction and to thus produce more marked effects.

The reflector is composed of a sheet of zinc 2 metres in height, curved and held on a wooden frame so as to have the form of a cylindrical surface having a parabolic section. The focal line is then parallel to the generatrices of the cylinder.

If the vibrator is placed so that its axis coincides with the focal line, the waves are propagated in the direction of the reflector's plane of symmetry and are perceptible by the resonator at a much greater distance than before. The distance at which they can be perceived is still further increased by placing the resonator along the focal line of a reflector similar to the first, so situated that the planes of symmetry of the two coincide.

An insulating screen, such as a wall, placed between the two parabolic *mirrors*, does not intercept the undulations; but a conducting screen stops them and casts a *shadow* behind it. It will be observed that we are naturally led to employ the language of optics to denote electric undulations.

These phenomena are identical with those produced by luminous rays; the only difference is in the order of magnitude of the wave-lengths; luminous waves are nearly a million times shorter than the electric waves produced by means of the vibrators used by the German scientist. The fact that Hertzian vibrations are not reflected by a wall may be compared to the fact that light is not reflected by a very thin transparent body: we know, indeed, that soap-bubbles no longer reflect the surrounding objects at the moment of their bursting; now the ratio of the length of light-waves to the thickness of the bubble is then of the same order as the ratio of the lengths of the Hertzian waves to the thickness of a wall.

Electric undulations are propagated in a straight line, as is shown by the fact of their stoppage by a metal screen. The mode of producing vibrations shows that they are parallel to the axis of the vibrator, *i.e.*, that they are transversal and, to use the corresponding optical expression, polarized rectilinearly.

If we place in the path of the rays transmitted by the first mirror a screen made of parallel wires, the effect of this screen is very different according as the wires are parallel or perpendicular to the axis of the vibrator. In the first case the electric waves pass without difficulty; in the second, the electric force is absorbed by the wires which are normal to it and the rays are extinguished. The effect is similar to that of a tourmaline plate in optics.

If, after having removed the screen, we turn the resonator and its mirror by an angle of 90° about the direction of the rays, no sparks are observed. But if the above-mentioned screen be interposed normally to the transmitted rays and inclining the wires 45° to the directions of the focal lines of the mirrors, the screen decomposes the incident waves and lets the vibrations inclined 45° to the axis of the resonator go past.

These can then act on the resonator. This phenomenon recalls the lighting up of the field of two crossed Nicol's prisms by the interposition of a crystal plate.

Lastly, the two mirrors enable us to exhibit clearly the phenomenon of the reflection and refraction of electric waves. For example, if we send an electric ray against a plane conductive partition, we can perceive the reflected waves by the aid of a resonator, provided that the planes of symmetry of the two parabolic mirrors be placed so that they intersect each other at the partition and that the normal plane through their intersection makes two equal plane angles.

To produce refraction Hertz made use of a large asphalt prism having a refringent angle of 30°. The incident rays directed upon the prism by the vibrator make with the refracted rays, which are received in the resonator, an angle indicating an index of refraction 1.7, only a slightly higher value than that given by optical experiments. Hertzian waves are therefore refracted by an insulating prism as light waves are by a glass prism.

These experiments by Hertz have been repeated by various physicists. MM. Sarrazin and de la Rive, of Geneva,* have shown that the form of the resonators can be modified without ceasing to perceive the sparks, and that the observed wave-length depends much more on the dimensions of this apparatus than on those of the exciter,—which seems to show that the latter gives rise to complex waves which can be picked out by suitable resonators. This explains why, when an oscillating discharge is kept up in a

---

* *Archives de Genève*, June, 1890.

room, sparks are seen to fly between metallic objects which are near together.

Lecher of Vienna has investigated the propagation of Hertzian waves along two parallel conductors, placing over their ends a Geissler tube to serve as a resonator. When a metallic bridge is slid along the conductors, the brilliancy of the tube grows less for certain positions of the bridge, and stronger for others. This phenomenon is comparable to the propagation of sound in a closed tube; at the nodes the pressure of the air is different from the atmospheric pressure, at the loops it is equal to this pressure. Consequently the sound is not altered if the tube be pierced at a loop, but it changes if the orifice is opposite a node. Thus at the nodes of Hertzian waves the difference of potential is maximum, and a metal bridge placed at these points over the two wires hinders the discharge from traversing the Geissler tube. If the bridge connects the wires at loops, its influence is null and the tube lights up. Lecher has found that the velocity of propagation of Hertzian waves in conductors is close to the velocity of light; Hertz had already found the same number for the velocity in air.

Hitherto we have only considered the electric field whose mean direction coincides with that of the axis of the vibrator. But the periodic currents of the oscillating discharge create, in addition, a magnetic field whose lines of force surround the conductors traversed by the undulatory electric flux (§ 135).

For a complete investigation of the phenomenon we must therefore examine at the same time both the effects of the electric forces propagated in form of waves, and of the perpendicular magnetic forces which accompany these disturbances in the surrounding medium, and whose effect are added to the first-mentioned.*

---

* For the details of Hertz's experiments see : Roosen, *Oscillations électri-*

224. Present Views on the Propagation of Electric Energy.—Hertz's experiments are a brilliant confirmation of the views, set forth by Faraday and defined by Maxwell, concerning the part played by the medium across which electric energy is transmitted.

It will be remembered (§ 109) that one of the hypotheses presented to account for electric discharge in conductors consists in supposing the latter to be the seat of a displacement of electricity comparable to the movements of fluid in pipes. Hence the expressions *electric current*, *flux of electricity*, and the various images borrowed from the dynamic theory of fluids in order to render the phenomena more easily understood at the outset.

But the investigation of the properties of the electric currents shows that there is a profound difference between it and fluxes of ponderable matter, in spite of the analogies met with in the laws governing these fluxes (§ 217).

When a fluid current circulates in a pipe, no external effect shows its presence ; the phenomenon is entirely concentrated inside the pipe itself.

The electric current, on the contrary, which makes itself evident in a conductor by giving out heat, exercises peculiar and very striking effects in the surrounding medium. It magnetizes the medium, as is shown by iron-filings figures ; it modifies the optic properties of bodies (§ 167), and, lastly, it produces induction-phenomena in conductors displaced in its field.

Bends in a pipe cause a loss of kinetic energy and diminish the blow given at the moment of stopping the flow ; the twisting of a conductor into spirals or a solenoid, on the

---

ques (Bull. Assoc. ing. sort. de l'Inst. Mont., 1890); Poincaré, *Électricité et Optique*, Paris, Carré, 1891 ; Lodge, *The Work of Hertz and some of his Successors*, Lond., 1894.

other hand, increases the energy of the extra-current on breaking the circuit.

A fluid current may be alternating; this is the case in a pipe where the sound-waves are propagated in the form of longitudinal vibrations. We are tempted to compare such a movement to an alternating electric current; but here, too, differences are evident, not only in the surrounding space, but even inside the conductor in which such currents pass. The sound-wave presents maximum displacements along the axis of the pipe, while alternating currents have their greatest strength towards the surface of the conductor (§ 208).

An electric current must be considered as the centre of a disturbance which affects all or part of the conductor by the Joule effect, and which extends out into the surrounding medium. As this propagation takes place in a vacuum, it follows that it is the ether that serves as vehicle for electric waves.

A current, at the moment it starts, excites an electromagnetic wave which is transmitted in the space surrounding the conductor with a velocity equal to that of light. When the current has reached its full strength, *i.e.*, when it has acquired a constant value in all the points of a section of the conductor, the surrounding medium is in a state of tension which is manifested by a tendency to contract in the direction of the magnetic lines of force and to dilate in a direction normal to them.

The ether about the conductor is then in a state of equilibrium characterized by cylindrical layers under a stress concentric to the conductor. When the current ceases, the ether, being suddenly un-stressed, falls back on the conductor, giving up to it its potential energy, which is shown in the extra-current.

An alternating current excites continuous waves, which as in the preceding case are propagated in space like luminous waves; the only difference lies in the duration of the period of the ether-vibrations.

It will be seen further on that alternating-current dynamos give from 50 to 200 vibrations per second. Given the enormous velocity of propagation in the ether, these currents produce waves several hundreds of kilometres long. The luminous vibrations of a lamp's flame amount to about 50 trillions per second, so that their wave-length is only some hundred-thousandths of a centimetre.

When these luminous radiations strike a body which intercepts them, it is found that their absorption causes a development of heat. So, too, conducting bodies which stop electric radiations are the seat of induced currents, made evident by calorific phenomena. The induced flux of electricity is in the same direction as the electric force and normal to the magnetic force of the wave.

This latter penetrates more or less into conductors, according as its length is greater or smaller. Short electromagnetic waves only affect the external layers of the conductors on which they fall. Hence the necessity of modifying the conductor's form, and adopting for short-period currents tubes, strips or ropes of metal in preference to solid conductors.

Electromagnetic waves, like radiations of heat and light, necessarily carry off with them a part of the energy from the source they leave to the conductors on which they impinge. To avoid this threatened loss in alternating-current circuits, the circuit is made of an outgoing and a return wire placed very close together. The waves emitted by the first go directly to the second, to which they restore, in the form of induced currents, the energy radiated off. This result is most completely obtained when two concentric conductors are used, a wire and a tube for example, for then the circuit has no action upon a neighboring magnet, and the magnetic field is rigorously limited to the space occupied by the conductors and the dielectric.

The discovery of the propagation of electromagnetic actions in the form of waves, similar to those of light, is of capital importance. It establishes an intimate bond between electricity, light, and heat, and will doubtless lead to important progress in the knowledge of the laws which govern these physical agents.

But so far only a corner has been lifted of the veil which hides the mechanism of the transmission of electric energy. We have learned that it is propagated without loss in dielectrics, while conductors are the seat of heat-effects which entirely or partly absorb the effective energy. But the phenomenon of the electric current still remains unexplained, even in its most simple form, viz., after the variable period is over.

From all modern experiments it is evident that an electric current is the manifestation of a transfer of energy which is taking place in the medium surrounding the conductors. These latter serve only to direct the propagation, and they perform this part at the expense of the absorption, under the form of heat, of part of the transmitted energy. A conductor should therefore be considered as the directrix along which the transfer takes place, just as the wick of a lamp is the centre of the flame without being the seat of the illuminating effect. The seat of the propagation of electric energy is in the electromagnetic whirls which encircle the conductors. As to the real mechanism of the transmission, it is as mysterious as the mechanism of gravitation.*

---

* Consult : Lodge, *Modern Views of Electricity ;* Stoletow, *Ether and Electricity* (*The Electrical World*, Jan. 28 et seq., 1893).

MEASUREMENT OF ELECTRIC POWER.

225. Case of a Continuous Current. — The electric power developed in a conductor, subjected to a potential-difference e and traversed by a current i, is expressed by the product ie.

This product can be determined indirectly by measuring the factors e and i separately, by means of one of the methods already given, or directly, by means of instruments called wattmeters.

226. Siemens Wattmeter. — This apparatus, which is applicable to the measurement of power developed by a continuous current, is made like the electrodynamometer, § 152, except that the circuits of the two coils are separate. The fixed coil, formed of a great number of turns of fine wire, is placed in shunt to the conductor in which the power to be measured is developed ; the moving coil, comprising a few turns of thick wire, is in series with this conductor and traversed by the same current as it.

The mutual action of the two coils is proportional to the product of the currents which traverse them ; but in consequence of the high resistance of the stationary coil it may be assumed, as with voltmeters, that the current traversing it is proportional to the original potential-difference e of the ends of the conductor. The electrodynamic couple, measured by the torsion-angle θ through which the suspension-spring must be turned in order to bring the movable coil back to its initial position, represents very nearly, therefore, the product ei; consequently

$$\theta = kei.$$

The factor k is determined by connecting the apparatus to a conductor of known resistance r, traversed by a current of known strength, α being then the angle of torsion,

$$k = \frac{\alpha}{i^2 r}.$$

227. Case of a Periodic Current.—When the electric energy is transmitted in the form of periodic currents, the choice of a method of measuring the power demands special attention.

There are then two cases to be considered :

I. The conductor in which is developed the power to be measured has a negligible self-induction; this is the case with straight or zigzagged wires and coils with special windings.

II. The self-induction is not negligible.

228. Non-inductive Conductors.—When a conductor of the first of the above classes is subjected to a periodic potential-difference, it becomes the seat of a current whose phases coincide with those of the potential-difference. Consequently, where the connection between the current and the time is a simple sine-curve, the mean power is equal to the product of the effective difference of potential by the effective current; for we then have, § 198,

$$\frac{1}{T}\int_0^T ei\,dt = \frac{E_\circ I}{T}\int_0^T \sin^2\frac{2\pi t}{T}\,dt = \frac{E_\circ I}{2} = \sqrt{(e^2)_m}\,\sqrt{(i^2)_m}.$$

The effective current can be determined separately by means of an electrodynamometer, or by a voltmeter having a negligible coefficient of self-induction—the Cardew volt-meter, for example.

If we have to deal with a complex periodic function, or one the form of which is unknown, the preceding method should not be used.

On the hypothesis that the power under consideration is completely transformed into heat by the Joule effect, we can then deduce the quantity of this heat from the current or potential-difference given respectively by the readings of an electrodynamometer or a Cardew voltmeter. Denoting by r the resistance, supposed to be known, of the conductor, we have

$$\frac{1}{T}\int_0^T ei\,\mathrm{d}t = r\frac{1}{T}\int_0^T i^2\,\mathrm{d}t = \frac{1}{r}\frac{1}{T}\int_0^T e^2\,\mathrm{d}t.$$

The Siemens wattmeter is not applicable in the case of periodic E. M. Fs. on account of the considerable self-induction of the fixed coil. The current produced in this coil would depend on· its impedance, that is, on the frequency of the periods. Zipernowski has sought to escape this difficulty by making each of the coils of a small number of turns, so as to render their self-induction feeble. At the end of the coil which is to be placed in shunt are placed supplementary coils, wound double, whose only function is to supply the total resistance desired in the shunt. Call this resistance ρ, r that of the conductor in which we are trying to find the power developed, and R the resistance of the wattmeter's second circuit. The connections of the apparatus are generally arranged so that the resistance ρ is in parallel to the two resistances r and R placed in series.

If the self-induction L_s of the branch ρ were zero, or, more exactly, if the *time-constant* $\dfrac{L_s}{\rho}$ were negligible, the current traversing this branch would be consonant in phase with the potential-difference e, and the reading of the apparatus would be directly proportional to the mean power

$$\theta = k\frac{1}{T}\int_0^T eidt,$$

the constant k being determined in the same manner as in the Siemens wattmeter.

But the time constant $\dfrac{L}{\rho}$ of the shunt is not negligible. The result of this is a diminishing of the current and a lag of the current-phase behind that of the potential-difference, which effects cause a reduction in the indications of the wattmeter below the value corresponding to the power actually developed in the conductor r.

As Mather has shown, the indications of the wattmeter may be rendered correct by winding the coil, which is put in series with the movable coil, in such a way that its capacity neutralizes the effect of the movable coil's self-induction.

229. Conductors having Self-induction.—The preceding methods are still at fault when treating a conductor whose self-induction is not negligible.

In this case there is a phase-lag between the current and the periodic E. M. F. which sets it up in the conductor, such that we have

$$\frac{1}{T}\int_0^T eidt < \sqrt{(e^2)_m}\ \sqrt{(i^2)}\ .$$

We have seen, for a simple periodic function that

$$\frac{1}{T}\int_0^T eidt = \sqrt{(e^2)}\ \sqrt{(i^2)_m}\ \cos\phi,$$

ϕ being the angle of lag.

If Zipernowski's modified wattmeter is applied to a conductor having self-induction, there is produced a retardation of phase both in the main circuit of the apparatus, in series with the conductor, and in the shunt. The result of this is a tendency towards consonance of phase in the two circuits, which may have the effect of raising the readings of the instrument beyond the value corresponding to the real power, contrary to what happens in the case of a conductor without self-induction.

This error is avoided if Mather's suggestion is utilized.

In the case under consideration we can arrive at rigorous results by the use of the calorimeter.

Fig. 104 gives a general idea of the calorimeter used by M. Roïti. The conductor TT', in which the power to be

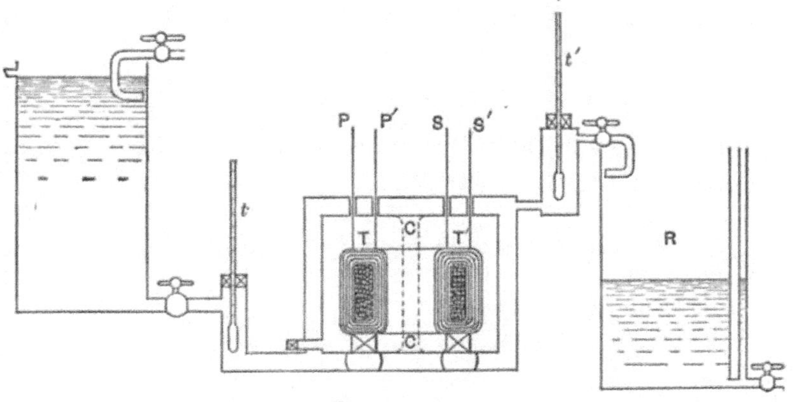

Fig. 104.

measured is developed and, by hypothesis, entirely transformed into heat, is enclosed in a brass vessel placed inside another vessel of the same metal: connecting wires enter through tubes. The space between the two vessels is traversed by a current of water which circulates under constant pressure and empties into a gauged receptacle R. Thermometers indicate, within a tenth of a degree, the temperature of the water on entering and leaving the double covering around the conductor. When the temperatures have become constant, which is sometimes a matter of many hours, the number of grammes of water flowing through per second and the difference in reading of the two-thermometers are noted. The product of these quantities multiplied by the mechanical equivalent of the gramme-degree is the power sought.

Messrs. Ayrton and Sumpner have worked out a method which allows of the exact determination of the energy consumed in an inductive resistance R, by means of a voltmeter and ammeter for alternating currents. A non-inductor resistance r is put in series with R, causing a fall of potential comparable with that caused by R. Let u_1 be the instantaneous potential-difference at the ends of R, u_2 that at the ends of r. The total fall of potential caused by $R + r$ will be at the given moment

$$u = u_1 + u_2. \quad \cdot \quad \cdot \quad \cdot \quad \cdot \quad \cdot \quad \cdot \quad \cdot \quad (1)$$

If a is the current at the time, the power at the given instant is

$$p = au_1 = \frac{u_2}{r} u_1.$$

But from (1) we deduce

$$u_1 u_2 = \tfrac{1}{2}(u^2 - u_1^2 - u_2^2),$$

whence

$$p = \frac{1}{2r}(u^2 - u_1^2 - u_2^2).$$

The mean power absorbed by R will be

$$P = \frac{1}{T}\int_0^T p\,\mathrm{d}t = \frac{1}{2rT}\left(\int_0^T u^2\mathrm{d}t - \int_0^T u_1^2\mathrm{d}t - \int_0^T u_2^2\mathrm{d}t\right)$$

$$= \frac{1}{2r}(U^2 - U_1^2 - U_2^2),$$

where U, U_1, U_2 represent the effective potential-differences at the ends of $R + r$, R and r measured with a Cardew voltmeter or an electrometer. If r is not known, it will suffice if the effective current A is found out by means of an electrodynamometer. We will then have, observing that $A = \dfrac{U_2}{r}$,

$$P = \frac{A}{2U_2}(U^2 - U_1^2 - U_2^2).$$

MEASUREMENT OF THE INTENSITY OF A MAGNETIC FIELD.

230. Method by Oscillation.—When a magnetic field may be considered uniform in the space in which a magnet-needle moves, the duration of a complete oscillation is expressed by

$$t = 2\pi \sqrt{\frac{\Sigma mr^2}{\mathfrak{M}\mathfrak{H}}},$$

Σmr^2 being the moment of inertia of the needle and $\mathfrak{M}\mathfrak{H}$ the product of the magnetic moment of the needle by the intensity of the field in which it moves. It is of course taken for granted that the field does not modify the magnetic moment of the needle, that is, that its intensity of magnetization corresponds to a greater magnetizing force than that of the field.

The above formula gives a simple method, either of comparing the intensity in different points of a field, or of determining an intensity in absolute value (§ 48).

231. Electromagnetic Method.—The author has worked out an apparatus which allows of measuring the intensity of a magnetic field by utilizing its action upon a conductor traversed by a known current. The instrument, shown in Fig. 105, was constructed by MM. F. Pescetto and P. Zunini when they were students at the Institut Montefiore. The conductor A, movable about an axis O and balanced by a counterweight P, is traversed for a portion l of its length by a current i led in by flexible wires f, f', and which can be measured in an ammeter.

If $\mathcal{H}$ is the component of the field-intensity normal to the plane of the conductor's displacement, the electromagnetic force acting upon this latter is

$$f = i\mathcal{H}l.$$

This force is counterbalanced by the elastic reaction of a spring-tongue R fixed on the movable conductor, and which is acted on by a micrometer-screw V. This screw is pre-

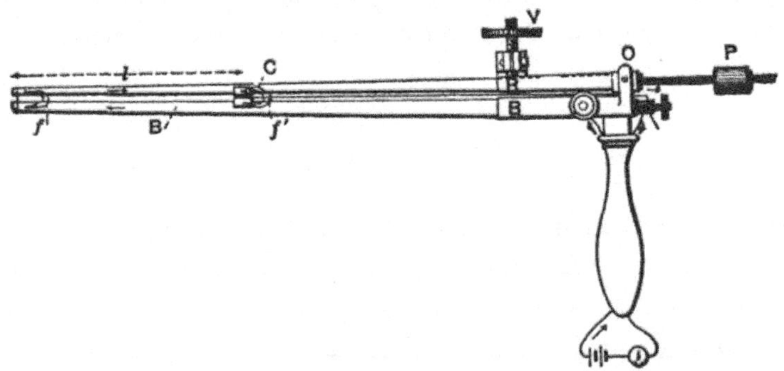

FIG. 105.

viously graduated by determining the weight which must be applied to the centre of the conductor l in order to balance the spring in its various states of tension and to bring the two arms A and B into parallelism. In consequence of the smallness of the electromagnetic reactions this apparatus is only suitable for measuring powerful fields, such as those in dynamo air-gaps, into which the flattened arms of the instrument can be introduced.

232. Method Based on Induction.—The method most generally employed is that described in § 192, which is based on induction. This method enables us to measure the intensity of very confined fields, such as that of a dynamo air-gap, for it is almost always possible to introduce a small flattened coil, connected to a ballistic galvanometer. If the coil be then sharply withdrawn to a position in which the flux of force traversing it is negligible, a displacement of electricity is produced in the circuit, which is measured by the angle of swing α of the galvanometer.

We have, § 150,

$$q = k \sin \tfrac{1}{2}\alpha.$$

But

$$q = \frac{\mathfrak{K}an}{R} \quad (\S\ 189),$$

whence

$$\mathfrak{K} = \frac{kR \sin \tfrac{1}{2}\alpha}{an}.$$

If the ballistic constant k of the galvanometer is unknown, it is determined by discharging into the apparatus a standard condenser charged by means of a cell of known E. M. F., or by introducing into the galvanometer-circuit an inclinometer (§ 192) which is made to turn in the earth's field so as to generate a calculable flux of electricity.

MEASUREMENT OF MAGNETIC PERMEABILITY.

233. Methods Based on Induction.—I. The apparatus shown in Fig. 106 has been used by Dr. Hopkinson. The

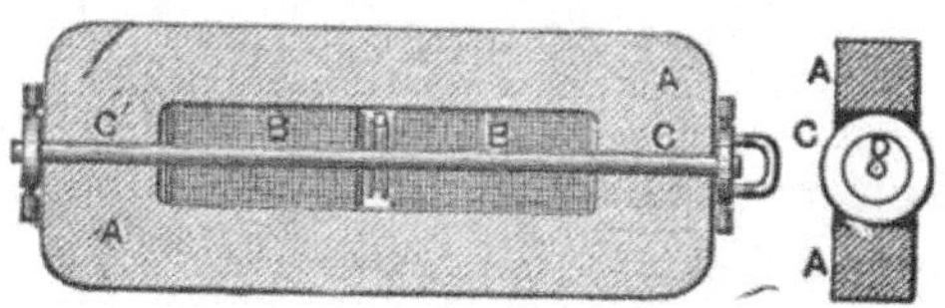

FIG. 106.

iron bar, whose permeability is to be measured, is formed of two sections, C and C'; it passes through a mass of wrought iron A, two magnetizing coils BB, and a small coil D connected with a ballistic galvanometer.

This coil D is pulled laterally by an elastic thread, so that if the two sections C and C' be slightly separated the coil is drawn out of the apparatus and traversed by an electric flux in proportion to the magnetic induction of the bar.

The iron mass and the bar form a magnetic circuit traversed by a flux produced by the magnetomotive force $4\pi nc$, n and c being respectively the total number of turns and the current of the coils BB.

Let Φ be the flux, l that length of the bar inside the open space in A, s its section, μ its permeability; also, let l' be the mean path of the lines of magnetic force in the mass A, s' its section, μ' its permeability.

We have (§ 164)

$$4\pi ni = \frac{\Phi l}{\mu s} + \frac{\Phi l'}{\mu' s'}.$$

In commercial tests, where a very close approximation is not called for, the second term of the binomial may be neglected in comparison with the first, and we have then simply

$$4\pi ni = \frac{\Phi l}{\mu s}. \qquad \cdots \qquad (1)$$

Now if n' denote the number of turns in the coil D, the electric flux produced by withdrawing this coil is

$$q = \frac{\Phi n'}{R} = k \sin \tfrac{1}{2}\alpha, \qquad \cdots \qquad (2)$$

R being the electric resistance of the circuit comprising the ballistic galvanometer.

By varying the magnetizing current c, measured in an ammeter, we can deduce from equations (1) and (2) the values of the permeability corresponding to different values of the magnetizing force. It is by means of such experiments that Dr. Hopkinson has been enabled to draw the continuous curves of Fig. 22, in which the permeability of various samples of iron is represented in terms of the magnetic induction.

234.—II. Fig. 107 shows an arrangement but slightly different from the preceding. The wrought-iron mass is not shown for the sake of clearness in the figure.

The bar B, surrounded by the mass of wrought iron, is undivided and fixed. It passes through the magnetizing coil M, traversed by a current measured in the ammeter a, and the coil S connected to the ballistic galvanometer G. This

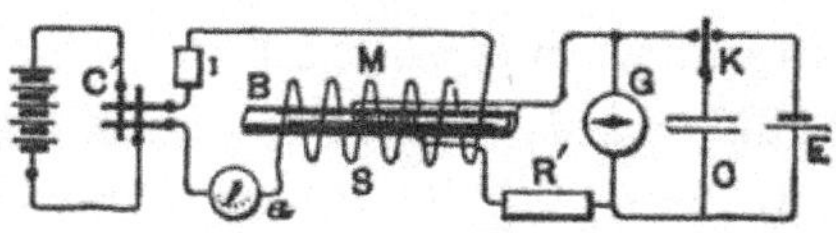

FIG. 107.

latter is graduated beforehand by means of a standard condenser C and a standard cell E, which enable us to determine the ballistic constant k.

When the direction of the current in M is changed by the key C', the flux which traverses the turns of the coil S varies by 2Φ.   Hence there is an electric flux in the galvanometer

$$q = \frac{2\Phi n'}{R} = k \sin \frac{\alpha}{2}. \qquad \ldots \quad \ldots \quad (1)$$

The rheostat R' serves to adjust the total resistance R so as to obtain suitable swings of the needle and to prevent the galvanometer from becoming dead-beat in the case where a Deprez-d'Arsonval instrument is used.

If, instead of reversing the magnetizing current, it is cut off, there remains a flux in the bar due to the residual magnetism of the system formed by the core and the enveloping mass. The electric flux q' measures in this case the difference between the total magnetic flux and the residual.

By adding to equation (1) the relation

$$4\pi n i = \Phi \frac{l}{\mu s}, \qquad \ldots \quad \ldots \quad \ldots \quad (2)$$

we can deduce the values of μ from the results of experiments.

To determine precisely the residual magnetism of the metal to be tested, Prof. Rowland gives the test-piece the form of a ring, the thickness of which along the axis is very much greater than the thickness normally to the axis. On completely surrounding such a ring with a magnetizing coil, a constant field is obtained within it, of intensity $\mathcal{K} = 4\pi n_1 i$ (§ 153). An auxiliary coil, wound round the ring and connected to a ballistic galvanometer, enables the total flux across the core to be measured. By this arrangement there is no need of an additional iron mass, and the residual magnetism of the core can be measured exactly; but the preparation and winding present difficulties.

235. In order to obtain by the induction-method the successive points of the curve representing the increases of intensity of magnetization corresponding to increases in the magnetizing force, Fig. 18, it is indispensable not to proceed by reversals or interruptions, but by gradual increments or decrements of the current in the magnetizing coil. The corresponding impulses observed in the ballistic galvanometer enable us to calculate the progressive increases of magnetic flux across the core and, consequently, the values of the magnetic induction and intensity of magnetization. After finding by this means the points corresponding to a complete cycle of the magnetizing force, we can determine the loss by hysteresis in the cycle. This method has the disadvantage of accumulating in the curves the successive errors committed in determining the separate points.

It would be possible to avoid this accumulation of errors by substituting a straight cylindrical core for the shapes proposed by Messrs. Hopkinson and Rowland, on condition of giving the bars or wires a length equal to 400 or 500 times their diameter, so that the magnetization could be considered uniform in the medial region of the bar when introduced into a magnetizing coil of similar length and section (§ 55). The coil connected to the ballistic galvanometer must then be wound outside the magnetizing coil and put near its middle portion. If this test-coil is then drawn sharply back, we will get an impulse in the galvanometer enabling us to calculate the total flux produced by the core and the magnetizing coil. To prevent the vibrations, caused by drawing back the test-coil, from being communicated to the core and affecting its magnetization, this coil may be wound on a glass tube surrounding the magnetizing coil.

236. Magnetometric Method.—Instead of treating the specimen-bar by the induction-method, its magnetic moment, plus that of the magnetizing coil, can be directly determined by the magnetometric method described in § 48. For this purpose the bar is placed in the plane of a suspended magnet-needle, and by the deviation of the needle we calculate the ratio of the total moment sought to the horizontal component of the earth's field. This latter, and also the effect of the magnetizing coil on the needle, are separately determined. For a straight electromagnet having the proportional dimensions indicated above, the poles are practically situated at the ends of the core. The electromagnet may be placed vertically; and in this case, if its length is sufficient and if one of the poles is very close to the needle, the effect of the other pole becomes negligible.

This proceeding gives us the magnetic moment and, consequently, the intensity of magnetization of the rod for cyclic values of the magnetizing force, whence we deduce the corresponding loss by hysteresis. The magnetometric method has the disadvantage of necessitating the determination of the horizontal component of the earth's field.

237. Method by Portative Power.—The portative power
of a magnet is expressed in terms of its magnetization by
the formula

$$p = 2\pi \mathfrak{J}^2 s. \quad (\S\ 56)$$

If the magnet be under the influence of a magnetic field,
we must add a second term, $\mathfrak{H}\mathfrak{J}s$, and then we have

$$p = 2\pi \mathfrak{J}^2 s + \mathfrak{H}\mathfrak{J}s. \quad \dots \quad \dots \quad (1)$$

Now

$$\mathfrak{J} = \kappa \mathfrak{H}, \quad (\S\ 52) \quad \dots \quad \dots \quad (2)$$

and

$$\mu = 1 + 4\pi\kappa. \quad \dots \quad \dots \quad (3)$$

Combining (1), (2), and (3), we obtain

$$\mu = \sqrt{\frac{8\pi p}{\mathfrak{H}^2 s} + 1}.$$

Now for a very long solenoid comprising n_1 turns per cm.
of length and traversed by a current i, $\mathfrak{H}$ is sensibly equal to
$4\pi n_1 i$.

MM. Mélotte and Henrard, students at the Liège Electrotechnical Institute, have utilized these properties in the

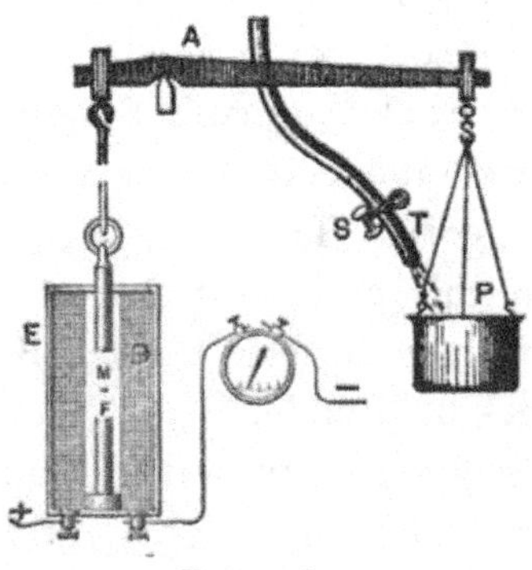

Fig. 108.

construction of a commercial apparatus for measuring permeabilities. The bar to be tested is divided into two parts,
M, F, placed in the middle of a long coil completely enclosed

in an iron sheath. The part M is hung from one of the ends of a balance-beam A. The other end of this beam carries a vessel P, into which sand is run through a tube T until the two parts of the bar are separated, against the magnetic attraction, produced by the passage of a current in B, which tends to hold them in contact.

The part M is slightly conical, and rests by its smaller base upon the larger end of F; in this way a well-defined surface of contact is obtained.

This method cannot give the values of the permeability corresponding to small values of the magnetizing force, nor yet the necessary data for calculating the loss by hysteresis; but for the values of permeability met with in dynamos, the results agree practically with those obtained by the preceding methods.

MEASUREMENT OF COEFFICIENTS OF INDUCTION.

238. Maxwell and Rayleigh's Method for Measuring a Coefficient of Self-induction in Terms of a Resistance. —The coil d, whose coefficient of self-induction is $L_{,}$, is inserted in the fourth arm of a Wheatstone bridge, Fig. 109, set up with a ballistic galvanometer and three series of non-

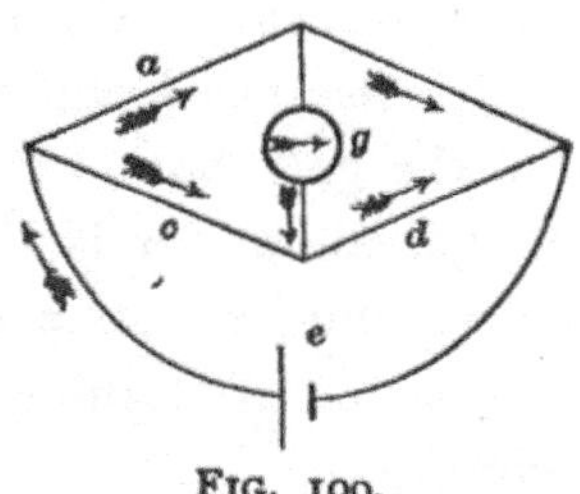

FIG. 109.

inductive resistances. These last are commonly formed of coils with double windings whose self-induction is not entirely negligible and which also possess an electrostatic capacity sufficient to cause sensible errors in the measurement of feeble coefficients of self-induction.　In such a case it is much better to employ very slender wires, straight or in a zigzag.

The four arms are first adjusted in the ordinary way on a constant current; next, the galvanometer arm is closed before that containing the battery; the reactions due to self-induction of the coil produce a deviation α.　When the battery circuit is opened, there is an equal and contrary deviation.

Denoting by D_1 the current, when fully established, in the arm d, the quantity of electricity conveyed by the extra-current on opening is (§ 194)

$$Q = \frac{L_s D_1}{d + b + \dfrac{g(a + i)}{g + a + i}};$$

the portion of this flux flowing through the galvanometer is

$$q = \frac{L_s D_1}{d + b + \dfrac{(a + i)g}{a + i + g}} \frac{a + i}{a + i + g} = k \sin \tfrac{1}{2}\alpha, \quad . \quad (1)$$

where k denotes the ballistic constant of the galvanometer (§ 150).

Lastly, the ohmic equilibrium of the bridge is destroyed by adding to d a small resistance r, which produces a permanent deflection δ in the galvanometer.

The current in the galvanometer is then approximately the same as if an electromotive force equal to rD_1 were acting in the arm d, the line containing the battery being interrupted.　We then have

$$G = \frac{r D_1}{d + b + \dfrac{(a + i)g}{a + i + g}} \frac{a + i}{a + i + g} = K\delta, \quad . \quad . \quad (2)$$

K being the reduction-factor of the galvanometer.

By combining equations (1) and (2) we deduce

$$L_s = r\frac{D_2}{D_1}\frac{k \sin\dfrac{\alpha}{r}}{K\delta}.$$

When a sensitive galvanometer is employed, the resistance to be added to d is very small, so that the currents D_1 and D_2 differ only by a negligible quantity, and their ratio is practically equal to unity; the formula then reduces to

$$L_s = r\frac{k}{K}\frac{\sin\frac{1}{2}\alpha}{\delta}.$$

The constants k and K are determined by preliminary experiments.

239. Maxwell's Method, modified by Pirani, for Measuring a Self-induction in Terms of a Capacity.—The relation existing between the effects of a self-induction and a capacity in a circuit traversed by a variable current (§ 219 *et seq.*) has suggested a number of methods for measuring one of these quantities in terms of the other. The following one is especially convenient.

A Wheatstone bridge (Fig. 110) has three non-inductive arms, a, b, s. In the fourth arm is placed the coil r, whose coefficient of self-induction is sought, in series with a non-inductive resistance r'. A condenser of capacity C is put in shunt on this latter.

First of all, the four arms are balanced for a fully established current; then they are balanced during the variable period by varying, for example, the point of connection of the condenser to the non-inductive resistance. Let R be the resistance of this fraction when the galvanometer remains at zero with the battery-circuit either open or closed. It was shown in § 219 that, in regard to the flux of electricity transmitted during the variable period, the condenser-effect corresponds to a decrease of self-induction equal to the product of the capacity by the square of the resistance of the conductor in parallel with the condenser.

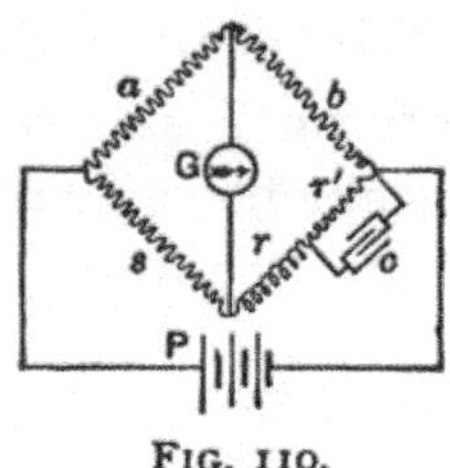

Fig. 110.

We shall then have in the present case

$$L_s = kR^2,$$

where k is the capacity of the condenser.

240. Ayrton and Perry's Method for Comparing a Coil's Self-induction with that of a Standard Coil.

—Ayrton and Perry's standard of self-induction enables the self-induction of a coil to be determined in the quickest possible way.

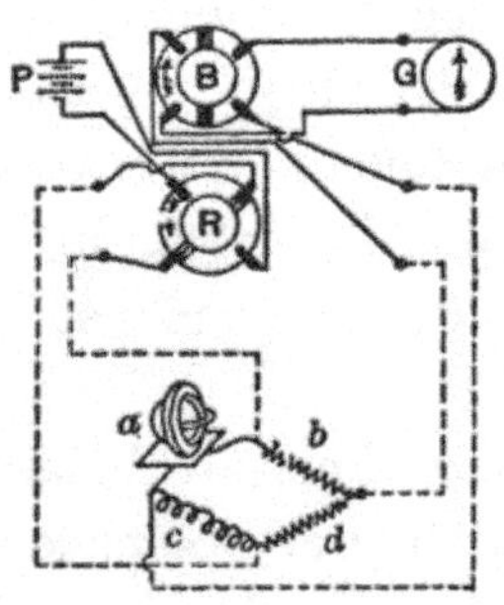

Fig. 111.

For this purpose a bridge is made (Fig. 111), by means of two non-inductive arms b, d, the coil to be standardized c, and the standard a. The galvanometer is brought to zero for a fully established current, then the battery-circuit is opened and closed while altering the position of the movable coil in the standard until the galvanometer remains at zero during the variable period.   We then have the proportion

$$\frac{r_b}{r_a} = \frac{L_{s_a}}{L_{s_c}}.$$

This method does not necessitate the long series of small adjustments needed by the other methods in order to obtain a balance in the two states (fully established and variable) of the current.   If the self-induction of the resistance c is beyond the bounds of the standard, it is only necessary to modify the ratio of the arms b and d, as is done in comparing, by Wheatstone's bridge, very different resistances.

Messrs. Ayrton and Perry have, moreover, designed a commutator which markedly increases the sensitiveness of both this and the preceding methods, by accumulating the effects on the galvanometer of the impulses due to repeated making and breaking of the battery-circuit.   For this purpose the connections of the bridge with the galvanometer and the battery are made by means of metallic brushes bearing on two movable discs B, R, composed of annular conducting segments separated by insulating spaces.   The movements of these discs are interdependent and adjusted so that the battery-circuit may be alternately open and closed, while the galvanometer is successively connected to the bridge and short-circuited.   The periods of connection of the galvanometer to the bridge correspond to the variable periods of opening and closing, and the connections are alternated in such a manner that the momentary currents traverse the galvanometer-coil in the same direction.   With this cumulative method a slight error in equilibrium is shown by a perceptible deflection of the needle.

Note.—When investigating an electromagnet by the preceding methods, the results found for the self-induction vary with the current which traverses the field-coils, since the coefficient of self-induction is a function of the permeability of the core (§ 188). It is therefore necessary to indicate the maximum value of the current traversing the field-coils for each value of the coefficient of self-induction.

241. Mutual Induction—Carey-Foster Method.—Suppose two concentric coils with or without an iron core; denote by L_m their mutual induction; insert one of them in a circuit through which a constant current i is sent; a flux of magnetic force $L_m i$ then traverses the second coil. If this latter is in communication with a ballistic galvanometer so as to form a circuit of total resistance R, the galvanometer is traversed by a flux of electricity, measured by a deflection α, when the current in the first coil is stopped. We have

$$\frac{L_m i}{R} = K \sin \tfrac{1}{2}\alpha. \qquad \ldots \ldots \quad (1)$$

A condenser of capacity c is next charged by connecting its plates to the extremities of a resistance traversed by the same current i. The discharge of the condenser into the same galvanometer gives a deflection α', such that

$$cir = K \sin \tfrac{1}{2}\alpha'. \qquad \ldots \ldots \quad (2)$$

Equations (1) and (2) give

$$L_m = crR \frac{\sin \tfrac{1}{2}\alpha}{\sin \tfrac{1}{2}\alpha'}.$$

Carey-Foster has given an arrangement enabling the two preceding combinations to be performed simultaneously so as to balance the effects of the two electric fluxes on the needle.   The galvanometer G, Fig. 112, is connected to the condenser and to the resistance r, on the one hand, and to the circuit of resistance R on the other.   By varying r until

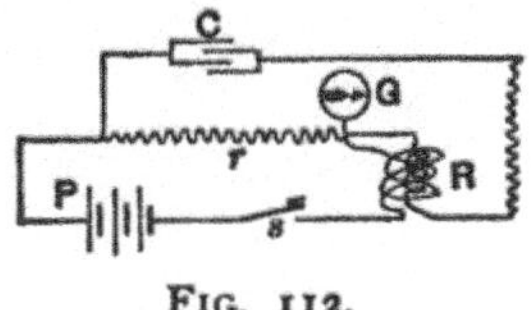

Fig. 112.

the needle remains at zero, both on making and breaking the battery-circuit, we have the relation

$$L_m = crR.$$

INDEX.